Reliability Engineering
Theory and Applications

Advanced Research in Reliability and System Assurance Engineering

Series Editor
Mangey Ram
*Professor, Assistant Dean (International Affairs)
Department of Mathematics; Computer Science & Engineering,
Graphic Era Deemed to be University, Dehradun, India*

Modeling and Simulation Based Analysis in Reliability Engineering
Mangey Ram

Reliability Engineering
Theory and Applications
Ilia Vonta and Mangey Ram

Reliability Engineering
Theory and Applications

Edited by
Ilia Vonta and Mangey Ram

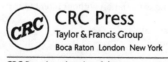

CRC Press is an imprint of the
Taylor & Francis Group, an **informa** business

CRC Press
Taylor & Francis Group
6000 Broken Sound Parkway NW, Suite 300
Boca Raton, FL 33487-2742

First issued in paperback 2021

© 2019 by Taylor & Francis Group, LLC
CRC Press is an imprint of Taylor & Francis Group, an Informa business

No claim to original U.S. Government works

ISBN-13: 978-0-367-78095-1 (pbk)
ISBN-13: 978-0-8153-5517-5 (hbk)

This book contains information obtained from authentic and highly regarded sources. Reasonable efforts have been made to publish reliable data and information, but the author and publisher cannot assume responsibility for the validity of all materials or the consequences of their use. The authors and publishers have attempted to trace the copyright holders of all material reproduced in this publication and apologize to copyright holders if permission to publish in this form has not been obtained. If any copyright material has not been acknowledged please write and let us know so we may rectify in any future reprint.

Except as permitted under U.S. Copyright Law, no part of this book may be reprinted, reproduced, transmitted, or utilized in any form by any electronic, mechanical, or other means, now known or hereafter invented, including photocopying, microfilming, and recording, or in any information storage or retrieval system, without written permission from the publishers.

For permission to photocopy or use material electronically from this work, please access www.copyright.com (http://www.copyright.com/) or contact the Copyright Clearance Center, Inc. (CCC), 222 Rosewood Drive, Danvers, MA 01923, 978-750-8400. CCC is a not-for-profit organization that provides licenses and registration for a variety of users. For organizations that have been granted a photocopy license by the CCC, a separate system of payment has been arranged.

Trademark Notice: Product or corporate names may be trademarks or registered trademarks, and are used only for identification and explanation without intent to infringe.

Visit the Taylor & Francis Web site at
http://www.taylorandfrancis.com

and the CRC Press Web site at
http://www.crcpress.com

Dedicated to Alex

Contents

Preface ix
Acknowledgments xi
Editors xiii
Contributors xv

1 **Optimal Maintenance for System with Two Failure Types** 1
   Stephanie Dietrich and Waltraud Kahle

2 **Availability Modeling of Complex Systems with Accelerated Deterioration Due to Fault Interactions: An Application in Traveling Wave Tubes** 17
   Stergios Tsiantis and Agapios N. Platis

3 **Modeling a Multi-State K-out-of-N: G System with Loss of Units** 43
   Ruiz-Castro Juan Eloy and Mohammed Dawabsha

4 **Reliability Analysis of Multi-State Cloud-RAID with Imperfect Element-Level Coverage** 61
   Lavanya Mandava, Liudong Xing, Vinod M. Vokkarane, and Ola Tannous

5 **L_z-Transform Approach for Comparison of Different Schemas of Ships' Diesel-Electric Multi-Power Source Traction Drives** 83
   Ilia Frenkel, Igor Bolvashenkov, Lev Khvatskin, and Anatoly Lisnianski

6 **Reliability Indicators for Hidden Markov Renewal Models** 105
   Irene Votsi

7 **Reliability Measures and Indices for Amusement Park Rides** 121
   Stavros Kioutsoukoustas, Alex Karagrigoriou, and Ilia Vonta

8 **On Parameter Dependence and Related Topics: The Impact of Jerzy Filus from Genesis to Recent Developments** 143
   Jerzy Filus, Lidia Filus, Barry C. Arnold, Pavlina K. Jordanova, Ludy Núñez Soza, Ying Lu, Silvia Stehlíková, and Milan Stehlík

9 **On Reliability of Renewable Systems** 173
   Vladimir Rykov

10 **Fuzzy Reliability of Systems Using Different Types
   of Level (λ, 1) Interval-Valued Fuzzy Numbers** 197
   Pawan Kumar and S. B. Singh

Index 215

Preface

Reliability theory is a multidisciplinary science aiming to develop complex technical and informational systems that are resistant to failures. During the last 50 years, numerous research studies have been published that focus on reliability engineering. Additional experience has also been gathered from industry. Thus, recently, reliability engineering has emerged as one of the main fields not only for theoretical scientists and researchers, but also for engineers and industrial managers.

This book covers the recent developments in reliability engineering. It presents new theoretical issues that were not previously presented in the literature, as well as the solutions of important practical problems and case studies illustrating the applications methodology.

The book is a collective work by a number of leading scientists, analysts, mathematicians, statisticians, and engineers who have been working on the front end of reliability science and engineering. All chapters in the book are written by leading researchers and practitioners in their respective fields of expertise and present various innovative methods, approaches, and solutions to various problems.

The editors would like to thank all the authors for accepting their invitation to contribute their work to this book, as well as all anonymous referees for an excellent job in reviewing the chapters and making their presentation the best possible.

Ilia Vonta
National Technical University of Athens, Greece

Mangey Ram
Graphic Era Deemed to be University, India

Acknowledgments

The editors acknowledge CRC Press for this opportunity and professional support. Our special thanks to Cindy Renee Carelli, Executive Editor, CRC Press—Taylor & Francis Group for the excellent support she provided to us to complete this book. Thanks to Renee Nakash, Editorial Assistant to Ms. Carelli for her follow-up and aid. Also, we would like to thank all the authors and reviewers for their availability for this work.

Ilia Vonta
National Technical University of Athens, Greece

Mangey Ram
Graphic Era Deemed to be University, India

Editors

Dr. Ilia Vonta received her Bachelor degree at the University of Patras, Greece (BSc in Mathematics), and her MA and PhD in Mathematical Statistics at the University of Maryland. She worked previously at the University of Cyprus, Department of Mathematics. Currently, she is an associate professor at the National Technical University of Athens, School of Applied Mathematical and Physical Sciences and the Hellenic Open University, School of Social Sciences. She held also short visiting positions at the Université René Descartes, Paris, France. She has published more than 30 research papers in international journals, while she has also edited the publication of a scientific book by Springer and has been guest editor of four international scientific journals. She is an editorial board member of four international journals. Her areas of specialization are survival analysis, biostatistics, semiparametric statistics, nonparametric statistics, reliability.

Dr. Mangey Ram received his PhD degree major in mathematics and minor in computer science from G. B. Pant University of Agriculture and Technology, Pantnagar, India, in 2008. He has been a faculty member for around 10 years and has taught several core courses in pure and applied mathematics at undergraduate, postgraduate, and doctorate levels. He is currently a professor at Graphic Era Deemed to be University, Dehradun, India. Before joining Graphic Era, he was a deputy manager (probationary officer) with Syndicate Bank for a short period. He is editor-in-chief of *International Journal of Mathematical, Engineering and Management Sciences* and guest editor and member of the editorial board of various journals. He is a regular reviewer for international journals, including those published by the IEEE, Elsevier, Springer, Emerald, John Wiley, Taylor & Francis, and many other publishers. He has published 125 research publications through the IEEE, Taylor & Francis, Springer, Elsevier, Emerald, World Scientific, and many other national and international journals of repute and has also presented his works at national and international conferences. His fields of research are reliability theory and applied mathematics. Dr. Ram is a senior member of the IEEE, a life member of the Operational Research Society of India, the Society for Reliability Engineering, Quality and Operations Management in India, the Indian Society of Industrial and Applied Mathematics, a member of the International Association of Engineers in Hong Kong and the Emerald Literati Network in the UK. He has been a member of the organizing committee of a number of international and national conferences, seminars, and workshops. He was conferred with the Young Scientist Award by the Uttarakhand State Council for Science and Technology, Dehradun, in 2009. He was awarded the Best Faculty Award in 2011 and, recently, the Research Excellence Award in 2015 for his significant contribution to academics and research at Graphic Era.

Contributors

Barry C. Arnold
Department of Statistics
University of California
Riverside, California

Igor Bolvashenkov
Institute of Energy Conversion Technology
Technical University of Munich
München, Germany

Juan Eloy Ruiz-Castro
Department of Statistics and Operational Research and IEMath-GR, Faculty of Science
University of Granada
Granada, Spain

Mohammed Dawabsha
Department of Statistics and Operational Research and IEMath-GR, Faculty of Science
University of Granada
Granada, Spain

Stephanie Dietrich
Institute for Mathematical Stochastics
Otto-von-Guericke University
Magdeburg, Germany

Jerzy Filus
Department of Mathematics and Computer Science
Oakton Community College
Des Plaines, Illinois

Lidia Filus
Department of Mathematics
Northeastern Illinois University
Chicago, Illinois

Ilia Frenkel
Center for Reliability and Risk Management,
Industrial Engineering and Management Department
SCE- Shamoon College of Engineering
Beersheba, Israel

Pavlina K. Jordanova
Faculty of Mathematics and Informatics
Shumen University
Shumen, Bulgaria

Waltraud Kahle
Institute for Mathematical Stochastics
Otto-von-Guericke University
Magdeburg, Germany

Alex Karagrigoriou
Department of Statistics and Actuarial -Financial Mathematics
University of the Aegean
Samos, Greece

Lev Khvatskin
Center for Reliability and Risk Management,
Industrial Engineering and Management Department
SCE- Shamoon College of Engineering
Beersheba, Israel

Stavros Kioutsoukoustas
Hellenic Open University
Patras, Greece

Pawan Kumar
Department of Mathematics,
 Statistics and Computer Science
G.B. Pant University of Agriculture
 & Technology Pantnagar
Uttarakhand, India

Anatoly Lisnianski
The Israel Electric Corporation
Haifa, Israel

Ying Lu
Biomedical Data Science
Stanford University
Palo Alto, California

Lavanya Mandava
Department of Electrical and
 Computer Engineering
University of Massachusetts
 Dartmouth
North Dartmouth, Massachusetts

Agapios N. Platis
Department of Financial and
 Management Engineering
University of the Aegean
Chios, Greece

Vladimir Rykov
Department of Probability and
 Statistics
Gubkin Russian State University of
 Oil and Gas
Moscow, Russia

S. B. Singh
Department of Mathematics,
 Statistics and Computer Science
G.B. Pant University of Agriculture
 & Technology Pantnagar
Uttarakhand, India

Ludy Núñez Soza
Departamento de Matemática
Facultad de Ciencias
Universidad de Tarapacá
Arica, Chile

Silvia Stehlíková
Institute of Statistics
University of Valparaíso
Valparaíso, Chile

Milan Stehlík
Institute of Applied Statistics and
 Linz Institute of Technology
Johannes Kepler University, Linz &
 Institute of Statistics
University of Valparaíso
Valparaíso, Chile

Ola Tannous
Department of Computer Science
Illinois Institute of Technology
Chicago, Illinois

Stergios Tsiantis
Hellenic Open University
Patras, Greece

Vinod M. Vokkarane
Department of Electrical and
 Computer Engineering
University of Massachusetts Lowell
Lowell, Massachusetts

Ilia Vonta
School of Applied Mathematical and Physical Sciences
National Technical University of Athens
Athens, Greece

Irene Votsi
Laboratoire Manceau de Mathématiques
Le Mans Université
Le Mans, France

Liudong Xing
Department of Electrical and Computer Engineering
University of Massachusetts Dartmouth
North Dartmouth, Massachusetts

1

Optimal Maintenance for System with Two Failure Types

Stephanie Dietrich and Waltraud Kahle

CONTENTS

1.1 Introduction 1
1.2 Maintenance Policy and Optimization Problem 2
1.3 Modeling the System 4
1.4 Example for Cost Optimal Maintenance 10
References 14

In this chapter, a repairable system with continuous lifetime distribution and two different failure types is studied. The two failure types are minor failures (type 1) and large failures (type 2). The minor ones can be repaired by minimal repair and the large ones can only be dealt with through replacement.

Furthermore, preventive maintenance (PM) is undertaken at predetermined periodic times with interval τ. It is assumed that these maintenance actions have a positive influence on the failure intensity; they adjust the virtual age of the system to some value v. The costs of PM may depend on v and the current age of the system.

Our aim is to find a cost optimal strategy for PM with respect to both v and τ, as well as with respect to the maximal number of PMs before a complete renewal of the system.

1.1 Introduction

In contrast to the most common maintenance models, the model in this chapter assumes that there are different kinds of failures, which cannot all be removed by a minimal repair. We consider a repairable system with continuous lifetime distribution and two different failure types. The modeling of the different failure types is analogous to Beichelt [2, 4], who is the pioneer in using different failure types in the modeling of maintenance models. The two failure types are minor failures (type 1) and large failures (type 2). The minor ones can be repaired by minimal repair and the large ones can only be dealt with through replacement. The repairs after a failure are called corrective maintenances (CMs). The idea can be illustrated by the example of a car. If, for instance, the car cannot be used because of a defective generator, this can easily be corrected through a minimal repair. But, after a serious traffic accident caused, for instance, by run-down tires or brakes, the roadworthiness of the car can surely not be achieved through a minimal repair.

To avoid such failures and accidents, it is necessary to carry out regular PM actions. A part of such maintenance actions is, among others, to repair or replace all broken and worn items that will soon give up working anyway. Therefore, it is supposed that the system undergoes both PM and CM actions. A periodic imperfect PM policy with finite planning horizon is used to model the occurrence of PM actions.

In the literature, there are multiple maintenance models that consider different failure types and there are many ways to distinguish between these failure types. Colosimo et al. [7], for instance, examine an imperfect maintenance model with two different failure types. Type A failures can be predicted before they happen by a visual inspection. In this way, the repair costs are smaller than the repair costs of type B failures, which cannot be predicted. Lin et al. [14] proposed a sequential imperfect PM model with two independent failure types. The difference here is that type I failures are maintainable failures and type II failures are non-maintainable. Hence, PM actions can reduce the failure rate of maintainable failures but cannot change the failure rate of non-maintainable failures. Based on that model, Zequeira and Bérenguer [20] and Castro [6] further assumed for periodic maintenance models that both failure types are dependent. Besides this, there are also models that use repairable and irreparable failure modes; for example, in Wang and Zhang [19]. Although there are many publications on systems with multiple types of failures, the imperfect maintenance model described in this chapter has not yet been discussed in the literature.

In reference [9], we have considered the same problem for discrete lifetime distributions where the behavior of the system is that of a multi-state system. For discrete lifetime distributions, the system can be modeled by a Markov chain, which is not the case for continuous distributions. In this chapter, we are using stochastic processes in continuous time to model the system.

1.2 Maintenance Policy and Optimization Problem

For further research, the following assumptions about the failure process are made:

1. Initially, a new repairable system is installed.
2. The two failure types occur independently of each other. Whenever a failure occurs, it is either a minor one (type 1) with probability $1-p$ or a large one (type 2) with probability p. Type 1 failures are removed by minimal repair and type 2 failures are dealt with through replacements.
3. The repair times are negligibly small.

We consider a periodic imperfect PM policy with finite planning horizon. In the periodic PM policy, the system is preventively maintained at fixed time intervals and repaired when there are intervening failures. The maintenance policy used in this chapter is a direct generalization of the one dealt with in Beichelt [3]. Similar maintenance policies were applied, for instance, in Nakagawa [15], Sheu, Lin, and Liao [18], and Zequeira and Bérenguer [20].

In particular, the following assumptions are made for the used maintenance policy.

1. The PM actions are imperfect in the sense that each PM action reduces the virtual age of the system to a constant virtual age of $v \geq 0$.
2. PM is performed at $v+\tau, v+2\tau, \ldots, v+(N-1)\tau$ with $\tau > 0$, $v \geq 0$, and $N \in \{1, \ldots, N_{max}\}$.
3. If no type 2 failure occurred in $[0, v+N\tau)$, the system is replaced preventively at $v+N\tau$.

Note that the restriction of N by N_{max} is appropriate, since systems have a finite useful life. Therefore, in our cost optimization problem, a predefined maximum number of PM actions N_{max} will be taken into account.

The abovementioned maintenance policy contains the age replacement policy and the minimal repair policy as special cases. The first one is obtained if $p = 1$ and $v = 0$. Then, the system is replaced at the time of failure or at age τ, whichever occurs first. If $p = 0$ and $v = 0$, we have the minimal repair policy, which means that the system is always replaced at age τ and failures that occur between the periodic replacements are removed through minimal repair.

Obviously, the life cycle of the system ends with a type 2 failure or at time $v+N\tau$. In both cases, the system is renewed and begins its work as a new one. If we want to minimize the average costs per time unit, we have to calculate the mean cycle length and the mean costs per cycle. These costs contain the cost c_R of a renewal at the end of a cycle, the costs c_M of all minimal repairs during the life of the system, and the costs c_{PM} of all PMs. In the next section, we will find the mean cycle length and the expected number of type 1 failures during a life cycle.

1.3 Modeling the System

First, consider a repairable system that has only one failure type and that is not preventively maintained. Then, let $(T_n)_{n\geq 1}$ be the random failure times of that system. Now, let us introduce the counting process $N = (N_t)_{t\geq 0}$, which counts the CM actions under the assumption that there are no PM actions. The intensity function of this counting process is denoted by $\lambda^N(t)$, $t \geq 0$. Note that if each failure is removed through minimal repair, the process $N = (N_t)_{t\geq 0}$ is an inhomogeneous Poisson process.

Now, suppose that the system is preventively maintained. Then, the random failure times are denoted by $(T_n^*)_{n\geq 1}$. The corresponding failure counting process is $N^* = (N_t^*)_{t\geq 0}$. Thus, the random variable N_t^*, where $t \geq 0$, is the number of failures until t and $\lambda^{N^*}(t)$, where $t \geq 0$, is the intensity function of N^*.

Finally, let us assume that the system that undergoes PM has different failure types. To construct the failure counting processes of both type 1 and type 2 failures, one has to thin out the process $N^* = (N_t^*)_{t\geq 0}$. This method is described, for example, in Belyaev and Kahle [5]. The random failure times $(T_n^*)_{n\geq 1}$ are a point process. Every realization t_n^* of these random variables is of type 2 with probability p and of type 1 with probability $1-p$. Let us define a new sequence $(\Delta_n)_{n\geq 1}$ of $P(\Delta_n = 1) = p$ and $P(\Delta_n = 0) = 1-p$ for $n \geq 1$. If $\Delta_n = 1$, the failure that occurs at time t_n^* is of type 2; otherwise, if $\Delta_n = 0$, the failure at time t_n^* is of type 1. If one divides the random variables $(T_n^*)_{n\geq 1}$ in accordance with the realization of $(\Delta_n)_{n\geq 1}$, this will produce two new point processes $(T_k^{*'})_{k\geq 1}$ and $(T_k^{*''})_{k\geq 1}$ of the type 1 and type 2 failure times, respectively. Note that if $N^* = (N_t^*)_{t\geq 0}$ is an inhomogeneous Poisson process with intensity function $\lambda^{N^*}(t)$, the corresponding stochastic processes $(N_t^{*'})_{t\geq 0}$ and $(N_t^{*''})_{t\geq 0}$ are also inhomogeneous Poisson processes with intensity functions $(1-p)\lambda^{N^*}(t)$ and $p\lambda^{N^*}(t)$, respectively [10, p. 134]. Therefore, it holds that [16, p. 36]

$$P(N_t^{*''} = n) = \frac{1}{n!}\left(\int_0^t p\lambda^{N^*}(x)dx\right)^n \exp\left(-\int_0^t p\lambda^{N^*}(x)dx\right) \qquad (1.1)$$

for $n = 1, 2, \ldots$. Let T_1 and T_1^* be the random times of the first failure of a repairable system with only one failure type without PM and with PM, respectively. If one takes into account two failure types, one has T_1' and T_1'' as the random times of the first occurrence of a type 1 or type 2 failure of a repairable system without PM, respectively. Analogously, $T_1^{*''}$ and $T_1^{*'''}$ are the random times of the first occurrence of a type 1 or type 2 failure of

Optimal Maintenance for System with Two Failure Types

a repairable system with consideration of PM actions, respectively. In our modeling, the random variable $T_1^{*\prime\prime}$ is very important because a type 2 failure ends a replacement cycle. In what follows, some properties of the distribution of $T_1^{*\prime\prime}$ are given.

Lemma 1 (Distribution function of $T_1^{\prime\prime}$). Suppose T_1 is the random time of the first failure of a repairable system without PM and no distinction in failure types. It is assumed that a failure is of type 1 with probability $1-p$ and of type 2 with probability p. Let $T_1^{\prime\prime}$ be the random time of the first type 2 failure of a repairable system without PM. Then, $T_1^{\prime\prime}$ has the following distribution function:

$$F^{T_1^{\prime\prime}}(t) = 1 - \exp\left(-\int_0^t p\lambda^N(x)dx\right), \quad \forall t \geq 0, \tag{1.2}$$

where $\lambda^N(\cdot)$ is the intensity function of the failure counting process $N = (N_t)_{t \geq 0}$.

Proof 1. Equation 1.2 follows immediately from the point that $N = (N_t)_{t \geq 0}$ is an inhomogeneous Poisson process with intensity function $\lambda^N(t)$; the corresponding stochastic processes $N' = (N'_t)_{t \geq 0}$ and $N'' = (N''_t)_{t \geq 0}$ are also inhomogeneous Poisson processes with intensity functions $p\lambda^N(t)$ and $(1-p)\lambda^N(t)$, respectively [10, p. 134]. Therefore, as long as no type 2 failure occurs, the processes N' and N'' are inhomogeneous Poisson processes and the random variable N''_t is Poisson distributed with mean $\Lambda^{N''}(t) = E(N''_t) = \int_0^t p\lambda^N(x)dx$.

Using this, Equation 1.2 can be derived as

$$F^{T_1^{\prime\prime}}(t) = P(T_1^{\prime\prime} \leq t) = P(N''_t \geq 1)$$

$$= 1 - P(N''_t = 0) = 1 - \exp\left(-\int_0^t p\lambda^N(x)dx\right), \quad \forall t \geq 0.$$

Theorem 1.1: (Distribution function of $T_1^{*\prime\prime}$).

Suppose $T_1^{\prime\prime}$ is the random time of the first type 2 failure of a repairable system without PM. Let $T_1^{*\prime\prime}$ be the random time of the first type 2 failure of a repairable system with PM, following PM policy from Section 1.2. Then, for the distribution function $F^{T_1^{*\prime\prime}}(t) = P(T_1^{*\prime\prime} \leq t)$, it holds that

$$F^{T_1^{*''}}(t) = \begin{cases} 0, \text{if } t < 0 \\ F^{T_1''}(t), \text{if } t \in [0, v+\tau) \\ F^{T_1''}(v+\tau) + \left(1 - F^{T_1''}(v+\tau)\right)\left(\dfrac{F^{T_1''}(t-\tau) - F^{T_1''}(v)}{1 - F^{T_1''}(v)}\right), \\ \quad \text{if } t \in [v+\tau, v+2\tau) \\ F^{T_1''}(v+\tau) + \left(F^{T_1''}(v+\tau) - F^{T_1''}(v)\right) \displaystyle\sum_{i=2}^{k}\left(\dfrac{1 - F^{T_1''}(v+\tau)}{1 - F^{T_1''}(v)}\right)^{i-1} \\ \quad + \left(\dfrac{1 - F^{T_1''}(v+\tau)}{1 - F^{T_1''}(v)}\right)^{k}\left(F^{T_1''}(t - k\tau) - F^{T_1''}(v)\right), \\ \quad \text{if } t \in [v+k\tau, v+\{k+1\}\tau), k \geq 2. \end{cases}$$

(1.3)

Proof 2. Since $T_1^{*''}$ is the random time of the first type 2 failure of a system with PM actions, it holds that $F^{T_1^{*''}}(t) = 0$ for $t < 0$. Using maintenance policy from Section 1.2, the first PM time is $v+\tau$. Before this point in time, a repairable system with PM actions is identical with an equivalent repairable system without PM. Hence,

$$F^{T_1^{*''}}(t) = P(T_1'' \leq t) = F^{T_1''}(t), \quad \forall t \in [0, v+\tau). \tag{1.4}$$

During the time between the first and the second PM action, that is, $t \in [v+\tau, v+2\tau)$, it holds that

$$\begin{aligned} F^{T_1^{*''}}(t) &= P(T_1'' \leq v+\tau) + P(T_1'' \geq v+\tau)P(T_1'' \leq t-\tau \mid T_1'' \geq v) \\ &= F^{T_1''}(v+\tau) + \left(1 - F^{T_1''}(v+\tau)\right)\left(\dfrac{F^{T_1''}(t-\tau) - F^{T_1''}(v)}{1 - F^{T_1''}(v)}\right). \end{aligned} \tag{1.5}$$

For $t \in [v+k\tau, v+\{k+1\}\tau)$, $k \geq 2$, it holds that

$$F^{T_1^{*''}}(t) = P(T_1'' \le v + \tau)$$

$$+ \sum_{i=2}^{k} P(T_1'' \ge v + \tau) P(T_1'' \ge v + \tau \mid T_1'' \ge v)^{i-2} P(T_1'' \le v + \tau \mid T_1'' \ge v)$$

$$+ P(T_1'' \ge v + \tau) P(T_1'' \ge v + \tau \mid T_1'' \ge v)^{k-1} P(T_1'' \le t - k\tau \mid T_1'' \ge v) \quad (1.6)$$

$$= F^{T_1''}(v+\tau) + \left(F^{T_1''}(v+\tau) - F^{T_1''}(v)\right) \sum_{i=2}^{k} \left(\frac{1 - F^{T_1''}(v+\tau)}{1 - F^{T_1''}(v)}\right)^{i-1}$$

$$+ \left(\frac{1 - F^{T_1''}(v+\tau)}{1 - F^{T_1''}(v)}\right)^{k} \left(F^{T_1''}(t - k\tau) - F^{T_1''}(v)\right).$$

Remark 1 (Intensity Function of $N^* = (N_t^*)_{t \ge 0}$). The intensity function of the counting process $N = (N_t^*)_{t \ge 0}$ is

$$\lambda^{N^*}(t) = \begin{cases} 0, & \text{if } t < 0 \\ \lambda^N(t), & \text{if } t \in [0, v+\tau) \\ \lambda^N(t - k\tau), & \text{if } t \in [v + k\tau, v + \{k+1\}\tau), \quad k = 1, \ldots, N-1 \end{cases} \quad (1.7)$$

where $\lambda^N(t)$ is the intensity function of the counting process $N = (N_t)_{t \ge 0}$, which counts the failures of a repairable system without PM.

Note that the intensity function of the counting process $N = (N_t)_{t \ge 0}$ is equal to the hazard function of the time to the first failure of a new system, that is, $\lambda^N(t) = h^{T_1}(t)$ for $t \ge 0$.

As mentioned before, the random cycle length, that is, the time between two replacements, is

$$L_{v,\tau,N} = \min\{T_1^{*''}, v + N\tau\} \begin{cases} < v + N\tau, & \text{with } P(T_1^{*''} \le v + N\tau) \\ = v + N\tau, & \text{with } 1 - P(T_1^{*''} \le v + N\tau) \end{cases}. \quad (1.8)$$

and its distribution function is given by

$$F^{L_{v,\tau,N}}(t) = P(L_{v,\tau,N} \le t) = \begin{cases} 0, & \text{if } t < 0 \\ F^{T_1^{*''}}(t), & \text{if } 0 \le t < v + N\tau. \\ 1, & \text{if } t \ge v + N\tau \end{cases} \quad (1.9)$$

Theorem 1.2: (Mean Cycle Length).

In the case of maintenance policy from Section 1.2, the mean cycle length is

$$E(L_{v,\tau,N}) = v + N\tau - \int_0^v F^{T_1''}(t)dt$$

$$- \tau F^{T_1''}(v+\tau) \sum_{i=1}^{N-1}(N-i)\left(\frac{1-F^{T_1''}(v+\tau)}{1-F^{T_1''}(v)}\right)^{i-1}$$

$$+ \tau F^{T_1''}(v) \sum_{i=1}^{N-1}(N-i)\left(\frac{1-F^{T_1''}(v+\tau)}{1-F^{T_1''}(v)}\right)^{i} \quad (1.10)$$

$$- \int_v^{v+\tau} F^{T_1''}(t)dt \sum_{i=1}^{N}\left(\frac{1-F^{T_1''}(v+\tau)}{1-F^{T_1''}(v)}\right)^{i-1}.$$

Proof 3. We get:

$$E(L_{v,\tau,N}) = \int_0^{v+N\tau} \left(1 - F^{T_1^{*''}}(t)\right) dt$$

$$= v + N\tau - \int_0^{v+N\tau} F^{T_1^{*''}}(t) dt. \quad (1.11)$$

After inserting Equation 1.3 and using some algebra, one obtains Equation 1.10 for the mean cycle length.

To compute a maintenance cost rate, it is necessary to determine the random number of minimal repairs during a replacement cycle.

Theorem 1.3: (Mean Number of Type 1 Failures in a Replacement Cycle).

Suppose $Z_{v+N\tau}$ is the random number of type 1 failures in the replacement cycle with length $min\{T_1^{*''}, v + N\tau\}$. Then, it holds that

$$E(Z_{v+N\tau}) = \int_0^{v+N\tau} \left(1 - F^{T_1^{*''}}(x)\right) \lambda^{N^*}(x) dx - F^{T_1^{*''}}(v+N\tau). \quad (1.12)$$

Proof 4. Let Z_t be the random number of type 1 failures up to time t. The expected value of Z_t can be computed with the help of the conditional expectation

$$E(Z_t) = E(Z_t \mid T_1^{*''} < t) P(T_1^{*''} < t) + E(Z_t \mid T_1^{*''} \geq t) P(T_1^{*''} \geq t). \quad (1.13)$$

The first summand represents the case that the replacement cycle is terminated by a type 2 failure and the second summand represents the case that a preventive replacement ends the replacement cycle.

First, we consider the probability of k type 1 failures up to time t given that at time t the first type 2 failure occurs. As mentioned before, the process of type 1 failures is a thinned Poisson process with intensity $(1-p)\lambda^{N^*}(t)$ where $\lambda^{N^*}(t)$ is the intensity of the failure process with PM. Therefore, we get

$$E(Z_t \mid T_1^{*''} = t) = \int_0^t (1-p)\lambda^{N^*}(x)dx. \quad (1.14)$$

For the first term in Equation 1.13, it holds that

$$E(Z_t \mid T_1^{*''} < t) = \frac{\int_0^t E(Z_x \mid T_1^{*''} = x)dF^{T_1^{*''}}(x)}{F^{T_1^{*''}}(t)} \quad (1.15)$$

and the second one can be rewritten as follows:

$$E(Z_t \mid T_1^{*''} \geq t) = E(Z_t \mid T_1^{*''} = t). \quad (1.16)$$

Inserting Equations 1.15 and 1.16 in the formula for the (unconditional) mean value of Z_t (Equation 1.13), using partial integration and substitution we obtain, after some algebra

$$E(Z_t) = \int_0^t \overline{F}^{T_1^{*''}}(x)d\Lambda^{N^*}(x) - F^{T_1^{*''}}(t) \quad (1.17)$$

In the following, the optimization criterion of interest will be the average maintenance cost per unit time. In the underlying maintenance policy, there are two scenarios.

The first one is that there is no type 2 failure up to the first preventive replacement at time $v+N\tau$. In this case, the replacement cycle has length $[0, v+N\tau]$. The total costs comprise the costs for minimal repairs during the replacement cycle, the costs for $N-1$ PM actions, and the costs for a preventive replacement at $v+N\tau$. The probability of the first scenario is $1-P(T_1^{*''} \leq v+N\tau)$.

The second scenario is that a type 2 failure occurs before the preventive replacement takes place at $v+N\tau$. In this case, the total costs consist of the costs for minimal repairs up to the time of the type 2 failure, the costs of the PM actions, and the costs of a replacement, because the type 2 failure can only be removed through a replacement. The probability of the second scenario is $P(T_1^{*''} \leq v+N\tau)$.

Definition 1 (Cost Optimization Problem). Let c_M denote the costs for a minimal repair, c_{PM} the costs of PM, and c_R the costs of a replacement. The average maintenance costs per unit time are

$$C(v,\tau,N) = \left(1 - P\left(T_1^{*''} \le v + N\tau\right)\right)$$

$$\cdot \frac{\left(c_M E(Z_{v+N\tau} \mid T_1^{*''} \ge v + N\tau) + (N-1)c_{PM} + c_R\right)}{E(L_{v,\tau,N})}$$

$$+ P\left(T_1^{*''} \le v + N\tau\right) \left(\frac{c_M E(Z_{v+N\tau} \mid T_1^{*''} < v + N\tau)}{E(L_{v,\tau,N})} \right.$$

$$\left. + \frac{\sum_{k=1}^{N-1} c_{PM} 1_{\{v+k\tau < E(T_1^{*''} \mid T_1^{*''} < v+N\tau)\}} + c_R}{E(L_{v,\tau,N})} \right) \quad (1.18)$$

$$= \frac{c_M E(Z_{v+N\tau}) + c_R}{E(L_{v,\tau,N})} + \bar{F}^{T^{*''}}(v+N\tau) \frac{(N-1)c_{PM}}{E(L_{v,\tau,N})}$$

$$+ F^{T_1^{*''}}(v+N\tau) \frac{\sum_{k=1}^{N-1} c_{PM} 1_{\{v+k\tau < E(T_1^{*''} \mid T_1^{*''} < v+N\tau)\}}}{E(L_{v,\tau,N})},$$

where $E(T_1^{*''} \mid T_1^{*''} < v + N\tau)$ is the expected time of the first type 2 failure under the condition that a type 2 failure ends the replacement cycle and $1_{\{\}}$ is the indicator function. The optimization problem then has the following form:

$$\min_{v \in [0,\infty),\, \tau \in (0,\infty),\, N \in \{1,\ldots,N_{\max}\}} C(v,\tau,N). \quad (1.19)$$

Note that the extreme case $\tau = 0$ is excluded from the optimization problem (Equation 1.19). However, the other extreme case of perfect PM actions, that is, $v = 0$, is still part of the cost optimization problem.

1.4 Example for Cost Optimal Maintenance

For an example, we have to specify the underlying lifetime distribution and costs of PM that depend on v and τ. In references [8] and [9], we introduced several possible cost functions. Here, we restrict ourselves to the function

$$c_{PM}(v,\tau) = c_R\left(1 - \xi(v,\tau)^\delta\right), \quad v,\tau \ge 0, v+\tau \ne 0, \delta > 0, c_R > 0, \quad (1.20)$$

where

$$\xi(v,\tau) = \frac{v}{v+\tau}. \tag{1.21}$$

can be interpreted as the degree of repair in a Kijima type 2 model. This cost function was introduced, for instance, in Kahle [11], [12], and [13]. It has the following properties:

- The costs of PM actions are bounded below and above. They are greater than zero and smaller than the costs of a replacement, that is, $0 \leq c_{PM}(v,\tau) \leq c_R$.
- The better the PM actions, that is, the smaller the value of v, the higher the PM costs. In the case of perfect repair, that is, $\xi(v,\tau) = 0$, implying that $v = 0$, the PM costs are $c_{PM}(v,\tau) = c_R$. The same holds if v tends to zero and δ does not at the same time tend to zero.
- The worse the PM actions and the lower the distance between PM actions, that is, the higher the value of v and the lower the value of τ, the lower the PM costs. In the case of minimal repair, that is, $\xi(v,\tau) = 1$, meaning that v tends to infinity or $\tau = 0$, there are no PM costs; that is, $c_{PM}(v,\tau) = 0$.
- If v and τ tend to infinity and δ does not tend to zero, the PM costs tend to the cost of a replacement, that is, $c_{PM}(v,\tau)$ tends to c_R.
- The lower the value of δ, the lower the PM costs. If $v \neq 0$ and δ tends to zero, the PM costs tend to zero also.

As a lifetime distribution, we consider the modified Weibull distribution (MWD) introduced by Sarhan and Zaindin [17]. It generalizes some most commonly used distributions in survival analysis, such as the exponential, Rayleigh (RD), linear failure rate (LFRD) and Weibull (WD) distributions. The cumulative distribution function of the MWD is

$$F(x;\alpha,\beta,\gamma) = 1 - \exp(-\alpha x - \beta x^{\gamma}), \quad \forall x \geq 0. \tag{1.22}$$

If $\gamma = 2$, the MWD (α,β,γ) becomes the LFRD with parameters α and β. By setting $\alpha = 0$ and $\gamma = 2$, we get the RD with parameter β. In the case of $\alpha = 0$, we obtain the WD with parameters β and γ. Finally, if $\beta = 0$, we obtain the exponential distribution with parameter α. Further, we consider the reduced modified Weibull distribution (RMWD), which was introduced by Almalki [1]. In contrast to the MWD, the RMWD allows increasing, decreasing, and even bathtub shapes of the hazard function. Its cumulative distribution function (CDF) is given by

$$F(x;\alpha,\beta,\gamma) = 1 - \exp(-\alpha\sqrt{x} - \beta\sqrt{x}\exp(\gamma x)), \quad \forall x \geq 0. \tag{1.23}$$

The main objective of this section is to compute the cost optimal values for ν, τ, and N. The parameters of all distributions are chosen so that the expectation is 5. The number of PM actions is restricted to 10, that is, $N_{max} = 11$, and hence the system will be preventively replaced at the latest after 10 PM actions. The probability that a failure is of type 2 is assumed to be 10%, that is, $p = 0.1$. The costs of corrective maintenance actions, that is, c_M and c_R, are assumed to be constant. No general statement whether the PM costs are always between c_M and c_R can be made. This depends on the individual values of ν, τ, and δ and the ratio c_M/c_R. Since the resulting average maintenance costs per unit time are not convex and the cost optimization problem (Equation 1.19) could not be solved analytically, complete enumeration is used to find the optimal maintenance strategies. The optimal maintenance strategies in this section are computed with the statistical computing software R and have an accuracy of two decimal places. It is, however, important to note that all values in the following table are computed based on a numerical integration routine and are therefore only approximative solutions.

In the following, the optimal values for ν and τ and the optimal number of PM actions before a preventive replacement takes place are computed for different cost ratios c_M/c_R and different values of δ.

Optimal values in the case where costs are proportional to the degree of repair are given in Table 1.1. Note that if the cost optimal N is 1, there are many cost optimal values of ν and τ, namely, the cost optimal values of ν and τ form a fixed sum.

The contour plot in Figure 1.1 shows the dependency of the average maintenance costs on ν and τ for the MWD ($\alpha = 0.03, \beta = 0.004335, \gamma = 3$), the ratio of costs $c_M/c_R = 2$, $p = 0.1$, $N = 10$, $\delta = 0.5$, and the cost function for PM (Equation 1.20).

The numerical results in Table 1.1 lead to the following conclusions:

1. If τ does not tend to zero, it holds that the higher the value of δ, the more expensive the PM actions. Therefore, with a rising value of δ it becomes cost optimal to carry out less PM; that is, N is decreasing.
2. The lower the costs of a renewal compared with the costs of a minimal repair, that is, the higher the ratio c_M/c_R, the less expensive the costs for PM compared with the costs of a minimal repair. Therefore, with a rising ratio c_M/c_R the cost optimal distance between PM actions becomes shorter, that is, τ is decreasing, and the PM actions become better, that is, ν is decreasing also. Hence, it is favorable to carry out good PM actions more often, instead of doing more minimal repairs.
3. The cost optimal values of ν for the RMWD are higher than for the other distributions because of the higher failure rate at lower ages. The cost optimal values of τ for the RMWD are also higher than for the other distributions. The reason for this is that the failure rate for this distribution remains at a relatively low level for a while before it starts to increase again.

TABLE 1.1

Optimal Values in the Case where Costs are Proportional to the Degree of Repair

	LFRD $\alpha=0.01$ $\beta=0.029$	RD $\beta=0.0314$	WD $\beta=0.0057$ $\gamma=3$	MWD $\alpha=0.03$ $\beta=0.004335$ $\gamma=3$	RMWD $\alpha=0.1$ $\beta=0.1746$ $\gamma=0.1$
$c_M/c_R=0.5$ $\delta=0.125$	$N=11$	$N=11$	$N=11$	$N=11$	$N=11$
	$\nu=0.84$	$\nu=0.81$	$\nu=0.84$	$\nu=0.93$	$\nu=2.18$
	$\tau=3.52$	$\tau=3.61$	$\tau=2.66$	$\tau=2.88$	$\tau=7.71$
$\delta=0.5$	$N=11$	$N=11$	$N=11$	$N=11$	$N=9$
	$\nu=0.85$	$\nu=0.81$	$\nu=1.07$	$\nu=1.27$	$\nu=2.11$
	$\tau=6.63$	$\tau=6.6$	$\tau=3.88$	$\tau=3.77$	$\tau=10.94$
$\delta=1$	$N=9$	$N=9$	$N=9$	$N=10$	$N=11$
	$\nu=0$	$\nu=0$	$\nu=0$	$\nu=0$	$\nu=0.57$
	$\tau=9.41$	$\tau=9.29$	$\tau=5.85$	$\tau=6.73$	$\tau=13.66$
$c_M/c_R=1\,\delta=0.125$	$N=11$	$N=11$	$N=11$	$N=11$	$N=10$
	$\nu=0.61$	$\nu=0.59$	$\nu=0.68$	$\nu=0.74$	$\nu=1.91$
	$\tau=2.48$	$\tau=2.4$	$\tau=2.15$	$\tau=2.36$	$\tau=5.59$
$\delta=0.5$	$N=9$	$N=11$	$N=10$	$N=10$	$N=11$
	$\nu=0.61$	$\nu=0.7$	$\nu=0.89$	$\nu=0.99$	$\nu=2.25$
	$\tau=4.87$	$\tau=3.97$	$\tau=2.9$	$\tau=3.14$	$\tau=7.65$
$\delta=1$	$N=4$	$N=4$	$N=8$	$N=11$	$N=9$
	$\nu=0$	$\nu=0$	$\nu=0$	$\nu=0$	$\nu=0.58$
	$\tau=6.23$	$\tau=6.37$	$\tau=4.58$	$\tau=5.02$	$\tau=12.21$
$c_M/c_R=2\,\delta=0.125$	$N=11$	$N=11$	$N=11$	$N=11$	$N=11$
	$\nu=0.43$	$\nu=0.42$	$\nu=0.54$	$\nu=0.59$	$\nu=1.83$
	$\tau=1.8$	$\tau=1.74$	$\tau=1.72$	$\tau=1.89$	$\tau=4.4$
$\delta=0.5$	$N=10$	$N=10$	$N=11$	$N=10$	$N=11$
	$\nu=0.51$	$\nu=0.49$	$\nu=0.74$	$\nu=0.78$	$\nu=2.09$
	$\tau=3.02$	$\tau=2.93$	$\tau=2.25$	$\tau=2.53$	$\tau=5.57$
$\delta=1$	$N=11$	$N=6$	$N=10$	$N=8$	$N=11$
	$\nu=0$	$\nu=0$	$\nu=0$	$\nu=0$	$\nu=0.93$
	$\tau=4.56$	$\tau=4.23$	$\tau=3.76$	$\tau=4.01$	$\tau=8.67$

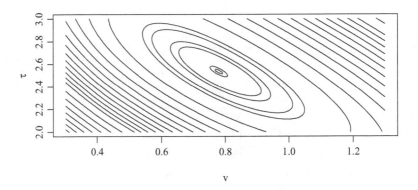

FIGURE 1.1
Average maintenance costs for the MWD $(\alpha = 0.03, \beta = 0.004335, \gamma = 3)$ if $N = 10$, $c_R = 500$, $c_M = 1000$, $p = 0.1$ and $\delta = 0.5$.

References

S. J. Almalki. A reduced new modified Weibull distribution. *arXiv:1307.3925*, 2013.

F. Beichelt. A general preventive maintenance policy. *Mathematische Operationsforschung und Statistik*, 7(6):927–932, 1976.

F. Beichelt. A generalized block-replacement policy. *IEEE Transactions on Reliability*, 30(2):171–172, 1981.

F. Beichelt. *Stochastic Processes in Science, Engineering, and Finance*. Chapman & Hall/CRC, Boca Raton, 1st edn, 2006.

Y. K. Belyaev and W. Kahle. *Methoden der Wahrscheinlichkeitsrechnung und Statistik bei der Analyse von Zuverlässigkeitsdaten*. B. G. Teubner, Stuttgart, 1st edn, 2000.

I. T. Castro. A model of imperfect preventive maintenance with dependent failure modes. *European Journal of Operational Research*, 196:217–224, 2009.

E. A. Colosimo, G. L. Gilardoni, W. B. Santos, and S. B. Motta. Optimal Maintenance Time for Repairable Systems Under Two Types of Failures. *Communications in Statistics: Theory and Methods*, 39(7):1289–1298, 2010.

S. Dietrich. Cost optimal maintenance in systems with imperfect maintenance. PhD thesis, Otto-von-Guericke-University Magdeburg, 2016.

S. Dietrich and W. Kahle. Optimal imperfect maintenance in a multi-state system with two failure types. In: I. Frenkel and A. Lisnianski, eds, *Proceedings of the Second International Symposium on Stochastic Models in Reliability Engineering, Life Science and Operations Management (SMRLO16)*, pp. 233–243. IEEE CPS, London, 2016. doi:10.1109/SMRLO.2016.47.

M. Finkelstein and J. H. Cha. *Stochastic Modelling for Reliability Shocks, Burn-In and Heterogeneous Populations*. Springer, London, 1st edn, 2013.

W. Kahle. Optimal incomplete maintenance for systems with discrete time-to-failure distribution. In: V. V. Rykov, N. Balakrishnan, and M. S. Nikulin, eds, *Mathematical and Statistical Models and Methods in Reliability*, vol. 1 of *Statistics for Industry and Technology*, pp. 123–132. Birkhäuser, Boston, 2011.

W. Kahle. Optimal incomplete maintenance in multi-state systems. In: A. Lisnianski and I. Frenkel, eds, *Recent Advances in System Reliability*, pp. 209–218. Springer-Verlag, London, 2012.

W. Kahle and E. Liebscher. *Zuverlässigkeitsanalyse und Qualitätssicherung*. Oldenbourg Wissenschaftsverlag, München, 1st edn, 2013.

D. Lin, M. J. Zuo, and R. C. M. Yam. Sequential imperfect preventive maintenance models with two categories of failure modes. *Naval Research Logistics*, 48(2):172–183, 2001.

T. Nakagawa. Periodic and sequential preventive maintenance policies. *Journal of Applied Probability*, 23:536–542, 1986.

S. E. Rigdon and A. P. Basu. *Statistical Methods for the Reliability of Repairable Systems*. John Wiley & Sons, New York, 1st edn, 2000.

A. M. Sarhan and M. Zaindin. Modified Weibull distribution. *Applied Sciences*, 11:123–136, 2009.

S. Sheu, B. Lin, and L. Liao. Optimal policies with decreasing probability of imperfect maintenance. *IEEE Transactions on Reliability*, 54(2):347–357, 2005.

G. J. Wang and Y. L. Zhang. Optimal repair-replacement policies for a system with two types of failures. *European Journal of Operational Research*, 226:500–506, 2013.

R. I. Zequeira and C. Bérenguer. Periodic imperfect preventive maintenance with two categories of competing failure modes. *Reliability Engineering and System Safety*, 91:460–468, 2006.

2

Availability Modeling of Complex Systems with Accelerated Deterioration Due to Fault Interactions: An Application in Traveling Wave Tubes

Stergios Tsiantis and Agapios N. Platis

CONTENTS

2.1 Introduction 17
2.2 Condition-Based Model of Complex Systems 20
    2.2.1 Description of the Proposed Model 20
    2.2.2 Semi-Markov Modeling 22
2.3 Availability Computation and Optimization 24
2.4 Application 24
    2.4.1 Factors Affecting TWT Deterioration 25
    2.4.2 TWT Availability Optimization 26
2.5 Conclusion 30
References 30
Appendix I 33
Appendix II 39

2.1 Introduction

Modern systems are complex and consist of many assemblies interacting with each other. Thus, a malfunction of one assembly can either cause complications in another or may accelerate its deterioration rate. According to Dekker et al. (1997), one type of dependency between components is stochastic dependence, where either the state of a component influences the lifetime distribution of others or there are external causes that induce simultaneous failures, which reduces the lifetime of others. For Liang and Parlikad (2015), there are two types of stochastic dependencies: inherent and induced. In systems suffering from inherent dependencies, the deterioration of their assemblies are affected by one another, while in induced dependence, malfunction of a component affects the deterioration of another. In complex systems, these interactions are more intensive. In some cases, a non-critical or less critical component's malfunction may accelerate the deterioration rate of a major component. This phenomenon is called *fault propagation*.

There is a large number of articles dealing with fault interactions; most attempt to specify the level where the malfunction of a component affects others and aim to optimize inspection intervals to prevent failures of major components. Rezaei (2015) and Golmakani and Moakedi (2012) investigated the optimization of inspection intervals in two component systems. Gao and Ge (2015) undertook a similar study considering the maintenance cost and three types of dependencies in components with two or three states. Zhang et al. (2014) studied maintenance techniques in complex assets where the malfunction of components affects the deterioration rate of others or induces sudden failures by using a modified iterative aggregation procedure (MIAP). Li et al. (2016) modeled stochastic dependencies of components using Levy copulas and suggested a new condition-based maintenance (CBM). Liang and Parlikad (2015) modeled CBM using continuous-time Markov chains (CTMCs) in complex assets with accelerated deterioration due to fault propagation. Sun et al. (2008) found that component dependencies increase the failure risk of recent repaired items. Lastly, Shi and Zeng (2016), Leung and Lai (2012), Sung et al. (2013), and Zhang et al. (2015) studied the optimization of preventive maintenance in multicomponent systems considering failure interactions.

To overcome these issues and increase system availability, preventive maintenance techniques are naturally performed, and there are two basic types of preventive maintenance approach: time interval maintenance, which is performed at specific time intervals and is intended to bring the system to an "as good as new" state (Dhillon 2002); and predictive maintenance, which is also called *CBM*. This last one is a very popular type of maintenance, whose philosophy stands on the idea that for the majority of failures, there are specific signs that indicate that a failure is imminent (Bloch and Geitner 2012). The system is inspected and the action to be taken depends on its deterioration state. There can be three types of actions taken: no action, minimal maintenance that recovers the system to the previous stage of deterioration, or major maintenance that brings the system to an "as good as new" state (Chen and Trivedi 2002). However, minimal or major maintenance may not always achieve its purpose and, in this case, they are considered imperfect. For example, major maintenance may sometimes not restore the system to an "as good as new state" but somewhere between "as good as new" and a previous deterioration state. According to Brown and Proschan (1983), during maintenance, either a system may not be fully repaired or the problematic part of a system may be repaired but a new failure to another part may be produced. Furthermore, Do et al. (2015a) indicate that imperfect maintenance is caused mainly by human factors such as the lack of knowledge and intention or by other factors such as the lack of spare parts, lack of time, or general maintenance policy for cost savings. In some cases, the imperfect maintenance/repair is due to the sensitivity of components.

CBM takes advantage of time interval maintenance because it uses all the data from the current state of the system (Ahmad and Kamaruddin 2012) and is more cost-effective because it is performed only when necessary, as it is triggered from the resulting inspections (Saranga 2002). In numerous papers modeling CBM, the fundamental objective is to determine the inspection intervals and the minimal and major maintenance thresholds to maximize availability and/or operational cost. Among the dominant approaches, there are CTMCs and Bayesian techniques and, in some cases, fault interactions are investigated as well. Chen and Trivedi (2002) provided closed form analytical results for CBM by modeling deterioration as a CTMC, considering also Poisson-type failures. Amari and McLaughlin (2004) studied the simple CBM model and provided a mathematical solution and an algorithm to maximize availability. Rao and Naikan (2006) undertook a similar investigation considering deterioration and sudden failures. Cekyay and Ozekici (2012) studied repair and replacement policies in systems that accomplish their mission in phases, where the deterioration depends on the phase of the mission. Chen and Trivedi (2005), again, investigated CBM policies using a semi-Markov decision process. Wang (2012) built a multivariate control chart based on Bayes theory, where maintenance policy depends on the comparison of *a posteriori* probability of the system being in the next state to a predetermined level. Caballé et al. (2015) took into account deterioration failures and sudden shocks, while Do et al. (2015a,b) modeled the deterioration process using the gamma process with perfect and imperfect maintenance. Lastly, Nguyen et al. (2015) modeled the deterioration process as a time-dependent stochastic process and studied CBM as a decision process at two levels: system and component.

Concerning the studies that take into account failure interactions, Liang and Parlikad (2015) created a CBM model for assets with accelerated deterioration due to fault propagation. They modeled the deterioration process as a multiple dependent deterioration path with multiple CTMCs. Tang et al. (2015) studied CBM using a semi-Markov process (SMP), considering also dependencies between deterioration and sudden failures. According to Amari et al. (2006), different deterioration mechanisms in a system should ideally be modeled separately. Rasmekomen and Parlikad (2016) studied optimization of CBM in complex systems with component dependencies. Their approach consists of three parts: independent degradation model, interactive degradation rate generic path (IDRGP) model using Gaussian process regression (GPR), and CBM optimization. Liu et al. (2013) studied CBM, considering multiple sudden shocks whose rate depends on the deterioration stage and the age of the system; and, finally, Do et al. (2015b) modeled CBM in two component systems with failure interactions, taking into account dependencies affected by their current state and economic dependencies.

In this chapter, a complex multicomponent system is considered. It consists of major/critical component and non-critical subcomponents. The major component, apart from its normal deterioration, suffers from accelerated deterioration due to malfunctions of its subcomponents. CBM is planned based on the inspection results to reduce the unexpected failures and extend the lifetime of the system. Another scope of CBM is to identify malfunctions of subcomponents that cause accelerated deterioration and eliminate their effects. Preventive maintenance can be minimal or major and both can be imperfect. To introduce a general model, it is assumed that the sojourn time at every state can be generally distributed, so the semi-Markov model is used to describe the system's deterioration over time. The purpose is to find the appropriate inspection intervals and the minimal and major maintenance thresholds that maximize availability of the system. The rest of the chapter is organized as follows: In Section 2.2, the CBM model of the system is described and the semi-Markov analysis is presented. In Section 2.3, the availability computation and optimization are presented. In Section 2.4, the proposed model is applied in the maintenance of military transmitters and the findings and conclusion of the chapter follow in Section 2.5.

2.2 Condition-Based Model of Complex Systems

2.2.1 Description of the Proposed Model

The proposed model is shown in Figure 2.1. Modeling of the deterioration is conducted in multiple paths, according to Liang and Parlikad's (2015) analysis. Except for normal deterioration of the major system, accelerated deterioration due to malfunctions in several subsystems is taken into account. The state of the system depends on the major and subsystem states and is characterized by two indices, a and b. Index a indicates the state of the major system, while index b shows the different types of subsystems' malfunctions. There are n types of subsystem malfunctions ($1 \leq b \leq n$). It is assumed that the joint probability of simultaneous transition of the state of the major system

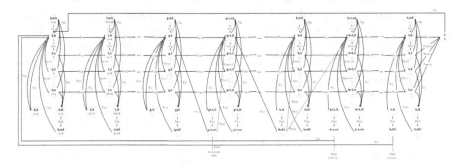

FIGURE 2.1
Condition-based maintenance model.

and malfunction of subsystems is negligible. If $b=0$, the system degrades at its normal deterioration rate F_i ($i=1, 2, ..., k$). If $1 \leq b \leq n$, the system degrades at an accelerated rate. For example, for $b=1$, the degradation rate follows the distributions F_{a1i} ($i=1, 2, ..., k$). The major system can transit from state $(a,0)$ to (a,b) at any time depending on the type of malfunction of the subsystems. The transition rate from states with $b=0$ to states with $b \neq 0$ depends on the distributions $F_{pr(ab)}$.

The major system is periodically inspected to identify its deterioration stage. After each inspection there can be no action, minimal maintenance, or major maintenance.

Minimal maintenance is implemented after stage g and major after stage b. Minimal maintenance returns the major system to its previous deterioration stage, while major restores the system to the "as good as new" state. At the last state, F (fail), corrective maintenance or general repair is performed.

Furthermore, both minimal and major maintenance can be imperfect. Thus, minimal imperfect maintenance can either fail to restore the system to the previous deterioration state or can lead the system to a worse deterioration state. Similarly, imperfect major maintenance can either fail to restore the system to the "as good as new" state and the system can remain at a state between the "as good as new" and fail states or can totally fail and cause complete failure of the system.

Subsystems are inspected also, and minimal maintenance is implemented to repair their malfunctions and return the major system to its normal degradation rate. Finally, the major system is subjected to sudden failures, which can appear at any state. Corrective maintenance is performed and restores the system to the state just before the failure. Sudden failures and state F are self-announced, while, for the remainder of the states, inspection is required.

Index b can also take the following values:

- i1: Major system inspection
- i2: Subsystem inspection
- m1: Major system maintenance (minimal or major)
- m2: Subsystem maintenance (minimal)
- m3: Sudden failure state

Under the assumption that the sojourn times at every state can be generally distributed, the SMP will be used to describe the system's evolution over time. Furthermore, due to the fact that the deterioration rate increases over time, two different inspection intervals will be considered for the major system and two for the subsystems. Specifically, until stage g, the major system is inspected at every T_1 time interval and subsystems at every T_3. After stage g, the major system is inspected at every T_2 time interval ($T_2 < T_1$) and subsystems at every T_4 ($T_4 < T_3$). All these inspection intervals are assumed to be fixed. Fixed time intervals are easier to use in practice, especially in complex systems in which different inspection times (following, e.g., exponential distribution) make maintenance programming too difficult.

2.2.2 Semi-Markov Modeling

Under the assumption that inspection intervals are fixed, they can be modeled by the unit step function $u(t-T)$, where:

$$u(t-T_i) = \begin{cases} 0, t < T_i \\ 1, t \geq T_i \end{cases}, i = 1,2$$

for the major system and

$$u(t-T_i) = \begin{cases} 0, t < T_i \\ 1, t \geq T_i \end{cases}, i = 3,4$$

for subcomponents.

Furthermore, in the proposed model (Figure 2.1), a red-colored number in every circle under the systems situation (a,b) indicates the number of each state. Supposing that the sojourn times at rest states are generally distributed, the evolution of the system over time can be described by an SMP $X(t)$, $t \geq 0$. Because the model has $N = (n+5)*g + (n+6)(k-g) + 1$ states, the process can be described by the following kernel matrix (Kulkarni 1995):

$$p(t) = \begin{bmatrix} p_{11} & p_{12} & \cdots & \cdots & p_{1N} \\ p_{21} & p_{22} & \cdots & \cdots & p_{2N} \\ \cdots & \cdots & \cdots & \cdots & \cdots \\ \cdots & \cdots & \cdots & \cdots & \cdots \\ p_{N1} & p_{N2} & \cdots & \cdots & p_{NN} \end{bmatrix}$$

The element $p_{ij}(t)$ of the matrix corresponds to the probability that if the SMP has just entered state i, it moves to the next state j within time t. This element can be defined according to Manatos et al.'s (2016) analysis, as follows:

$$p_{ij}(t) = \begin{cases} 0 \\ F_{ij}(t) \\ \int_0^t \overline{F}_{ik}(t) \overline{F}_{im}(t) F_{ij}(t) dt \end{cases} \quad (2.1)$$

where $F_{ij}(t)$ is the cumulative density function (cdf) of time spent in state i before moving to state j and $\overline{F}_{ik} = 1 - F_{ik}(t)$, with $F_{ik}(t)$ the cdf of time spent in state i before moving to state k.

It is clear that $p_{ij}(t) = 0$ when the SMP cannot be moved to any state from state i within time t, $p_{ij}(t) = F_{ij}(t)$ when the SMP can be moved only to state j from state i within time t and $p_{ij}(t) = \int_0^t \overline{F}_{ik}(t)\overline{F}_{im}(t)F_{ij}(t)dt$ when there is more than one state (e.g., k, m, j) to which the SMP can be moved from state i within time t.

According to the analysis of Chen and Trivedi (2005), Kulkarni (1995), and Trivedi (2002), one way to describe the SMP is to assume that the transition happens in two stages. In the first stage, the process remains at state i for time t with cdf $H_i(t)$, the sojourn time distribution in state i. In the second stage, the process transits from state i to state j with probability p_{ij}. By this two-stage method, the SMP is described by the one-step probability matrix P and the vector of sojourn times distribution $H(t)$. The one-step probability matrix is given by the limit of $p(t)$ as t tends to infinity: $p(\infty) = \lim_{t \to \infty} p(t)$. Furthermore, if the matrix $v = [v_1, v_2, ..., v_N]$ denotes the state probabilities of the embedded Markov chain, then to count the steady-state probabilities, the following equations should be solved: $v = vP(\infty)$, $\sum_{i=1}^{N} v_i = 1$. Moreover, the mean sojourn time h_i at every state i is given by:

$$h_i = \int_0^{\infty} [1 - H_i(t)]dt \qquad (2.2)$$

where $H_i(t)$ is the cdf of sojourn times at the states of the chain. If there is more than one possible state to which the SMP can be moved from state i, then the cdf of sojourn times is calculated as follows.

Let us suppose that there are two possible states j and k that the SMP can be moved from state i. If k_{ij} is the time spent at state i given that the next state will be state j and, correspondingly, k_{ik} is the time spent at state i given that the next state will be state k, then the cdf of the sojourn time at state i will be the minimum of the distributions of the two possible transitions:

$$H_i(t) = \Pr\left[\min(k_{ij}, k_{ik}) \leq t\right]$$
$$= 1 - \Pr[\min(k_{ij}, k_{ik}) > t] = 1 - \Pr(k_{ij} > t)\Pr(k_{ik} > t)$$
$$= 1 - \overline{F}_{k_{ij}} \overline{F}_{k_{ik}} \qquad \text{(Trivedi 2002).}$$

Finally, the probability of state i is given by the equation

$$\pi_i = \frac{v_i h_i}{\sum_j v_i h_i}. \qquad (2.3)$$

2.3 Availability Computation and Optimization

The aim of using the condition-based model of this chapter is to suggest the appropriate maintenance policy that maximizes the availability of the system and count the corresponding inspection times for the major system and subsystems. Thus, the designer of a complex system could plan the appropriate inspection intervals and the appropriate stages to perform minimal and major maintenance.

Availability is the probability that the system works under certain conditions, if the resources are available. If we separate the state space V of the proposed model into two subsets A and B, where A defines the operational states and B the nonoperational states, then it is clear that $V = A \cup B, A \cap B = \emptyset, A \neq \emptyset, B \neq \emptyset$. Subset B includes all the failure states, the inspection states, state F, and maintenance states. Subset A includes all the rest states.

Availability, which includes all the operational states, will be equal to:

$$\text{Availability} = \sum_{i \in A} \pi_i = \pi_1 + \pi_2 + \pi_3 + \ldots + \pi_{n+1} + \pi_{n+2} + \pi_{n+3} + \pi_{n+4}$$
$$+ \pi_{n+5} + \pi_{n+6} + \pi_{n+7} + \pi_{n+8} + \ldots \qquad (2.4)$$
$$+ \pi_{2n+6} + \pi_{2n+7} + \pi_{2n+8} + \pi_{2n+9} + \pi_{2n+10} + \ldots$$

Supposing that a system works for 8760 hours per year, then, to determine its availability, the following equation should be solved:

$$\max_{T_1,T_2,T_3,T_4} \text{Availability}(T_1, T_2, T_3, T_4) = \max_{T_1,T_2,T_3,T_4} \sum_{i \in A} \pi_i \qquad (2.5)$$

$$0 < T_1 < T_2 < 8760, \ 0 < T_3 < T_4 < 8760.$$

2.4 Application

In this section, the applicability of the model is investigated by applying it to the maintenance of traveling wave tubes (TWTs). The TWT is the main transmitter part of military radar. It amplifies the transmitted radiation of a radar unit. It consists of an electron gun, an anode and cathode, a heater, a grid, a helix, magnets, and a collector. Transmitters also consist of some subsystems such as the high voltage power supply (HVPS) and the cooling system. Both these subsystems affect the deterioration rate of the TWT. The

TWT has a limited useful operational time that depends on environmental and operational conditions. When it fails, major circuit replacement can be applied to restore it to an "as good as new" state.

2.4.1 Factors Affecting TWT Deterioration

The most critical factors that affect the TWT's deterioration rate are voltage, temperature, and cathode current. These factors affect the deterioration rate of the cathode, which is strongly related to the deterioration of the TWT. According to Sivan (2002), degradation of the cathode is accelerated by unusual conditions such as a false power supply. Thelander (2015) indicates that correct adjustment of the cathode current eliminates the degradation of the TWT in its first six years of operation and reduces its degradation rate in its remaining years of operation. As a result, the useful life of a TWT is increased by two or three years. Furthermore, an appropriate power supply to the cathode heater results in 30% to 50% longer TWT life (Communications and Power Industries). TWT degradation is also affected by the beam current, which depends on cathode and grid voltage. Even a small deviation from the required beam current can increase body current, which results in the cathode temperature increasing and acceleration of its degradation. Finally, an increase in grid voltage causes overheating due to secondary electron emission from the grid, which may cause cathode poisoning (Sivan 2002).

The correct power supply is strongly related to TWT reliability (Advisory Group for Aerospace Research & Development 1994). Thus, to avoid acceleration of the TWT deterioration rate, the power supply should be periodically checked and remedial actions such as the use of an oscilloscope to check the power at specific transmitter points and the adjustment of the cathode current based on test points, not on vector values, should be performed.

Cooling of the TWT is also a crucial factor, because overheating increases its deterioration rate. It is well known that older coolants' coolanol 25R hydroscopic composition caused catastrophic problems in F-16 aircraft radar APG-68 (Acton 2013). Thus, this coolant has been replaced by the more stable polyalphaolefins (PAO). PAO has improved resistance and stability compared with those of coolanol, but its performance also varies and depends on the liquid's flow and its viscosity. Specifically, its performance is affected by environmental and operational conditions (Timofeeva 2001). For example, PAO may be contaminated by humidity absorption and particles (Chiesa and Das 2009) and, as a result, its thermal conductivity may be reduced, which further decreases its dielectric strength (Khodorkovsky et al. 1997). Furthermore, the liquid's performance is reduced due to temperature increases, which affect its thermal stability (Fei et al. 2015; Nelson 2007). According to Bazan (2010), an increase in temperature of 50°C reduces its thermal conductivity by 10%. Moreover, Khodorkovsky et al. (1997) declare that an increase in temperature of 10°C doubles its degradation rate. Finally, according to Mudawar (2001), PAO's performance drops in temperatures that are too low because of the change in its viscosity.

PAO's variance in performance may cause inadequate cooling of the TWT, which results in an increase of its deterioration rate (Sivan 2002). Furthermore, inadequate cooling can cause magnets to overheat, which results in degassing of the TWT. The collector is also affected by inadequate cooling (Advisory Group for Aerospace Research & Development 1994). To overcome these problems, periodic inspection of the cooling system is necessary. This inspection may consist of a pressure check, dielectric strength check, and so on. Specific devices such as a PAO purifier are used to fulfill the cooling system in cases of leakage, dielectric strength check, and the cleaning of particles by the liquid.

2.4.2 TWT Availability Optimization

To model TWT deterioration, the proposed model in Figure 2.1 will be used, with $k=3$. The values of the deterioration parameters and sudden failures come from the experience of the author in the maintenance of TWTs in military radar systems. The rest of the parameters such as the inspection times and the maintenance distributions come from the literature. All the parameters are presented in Table 2.1. To model the deterioration process, different exponential distributions between the deterioration stages (F_i) are used. Exponential distribution is also used for sudden failures (F_p). The TWT undergoes two types of accelerated deterioration. The first is caused by cooling problems (F_{a1}) and the second by power supply problems (F_{a2}). The probability of moving from the normal deterioration path to the two accelerated paths is described by the exponential distributions F_{pr1} and F_{pr2}, respectively. This probability is strongly related to environmental and operational conditions and may differ for different systems and conditions. The time to carry out inspections is modeled by the unit step function F_{I1} for the TWT and F_{I2} for the subsystems. All inspection times have a fixed duration (Chen and Trivedi 2001). The duration of minimal and major maintenance and the repair time are exponentially distributed also (Chen and Trivedi 2001; Batun and Azizoğlu 2009). Furthermore, the exponential distribution describes imperfect maintenance due to its nature, which is mainly caused by random human errors.

The scope of this chapter is the study of TWT availability, which correlates with radar transmitter availability. Another objective is to determine the minimal and major maintenance thresholds (g and b) that maximize availability and count the appropriate number of inspections. Thus, the following cases will be studied:

- Case 1: $g=0, b=2$
- Case 2: $g=0, b=1$
- Case 3: $g=1, b=2$

Availability Modeling of Complex Systems

TABLE 2.1

Distribution of System States Sojourn Times

Distribution	Parameter Value	Distribution	Parameter Value
$F_1(t) = 1 - e^{-\lambda_1 t}$	$\lambda_1 = 0.000075$	$F_{pr2}(t) = 1 - e^{-\lambda_{pr2} t}$	$\lambda_{pr2} = 0.002$
$F_2(t) = 1 - e^{-\lambda_2 t}$	$\lambda_2 = 0.00027$	$F_p(t) = 1 - e^{-\lambda_p t}$	$\lambda_p = 0.00014$
$F_3(t) = 1 - e^{-\lambda_3 t}$	$\lambda_3 = 0.00075$	$F_{mp}(t) = 1 - e^{-\lambda_{mp} t}$	$\lambda_{mp} = 0.025$
$F_4(t) = 1 - e^{-\lambda_4 t}$	$\lambda_4 = 0.0012$	$F_{m1}(t) = 1 - e^{-\lambda_{m1} t}$	$\lambda_{m1} = 0.5$
$F_{a11}(t) = 1 - e^{-\lambda_{a11} t}$	$\lambda_{a11} = 0.0001$	$F_{im1}(t) = 1 - e^{-\lambda_{im1} t}$	$\lambda_{im1} = 0.01$
$F_{a12}(t) = 1 - e^{-\lambda_{a12} t}$	$\lambda_{a12} = 0.00035$	$F_{Fm1}(t) = 1 - e^{-\lambda_{Fm1} t}$	$\lambda_{Fm1} = 0.001$
$F_{a13}(t) = 1 - e^{-\lambda_{a13} t}$	$\lambda_{a13} = 0.001$	$F_{m2}(t) = 1 - e^{-\lambda_{m2} t}$	$\lambda_{m2} = 0.9$
$F_{a14}(t) = 1 - e^{-\lambda_{a14} t}$	$\lambda_{a14} = 0.0016$	$F_M(t) = 1 - e^{-\lambda_M t}$	$\lambda_M = 0.2$
$F_{a21}(t) = 1 - e^{-\lambda_{a21} t}$	$\lambda_{a21} = 0.0001$	$F_{IM1}(t) = 1 - e^{-\lambda_{IM1} t}$	$\lambda_{IM1} = 0.03$
$F_{a22}(t) = 1 - e^{-\lambda_{a22} t}$	$\lambda_{a22} = 0.00038$	$F_{IM2}(t) = 1 - e^{-\lambda_{IM2} t}$	$\lambda_{IM2} = 0.02$
$F_{a23}(t) = 1 - e^{-\lambda_{a23} t}$	$\lambda_{a23} = 0.00105$	$F_{IM3}(t) = 1 - e^{-\lambda_{IM3} t}$	$\lambda_{IM3} = 0.01$
$F_{a24}(t) = 1 - e^{-\lambda_{a24} t}$	$\lambda_{a24} = 0.0017$	$F_{FM}(t) = 1 - e^{-\lambda_{FM} t}$	$\lambda_{FM} = 0.002$
$F_{pr1}(t) = 1 - e^{-\lambda_{pr1} t}$	$\lambda_{pr1} = 0.003$	$F_R(t) = 1 - e^{-\lambda_R t}$	$\lambda_R = 0.002$
$F_{I1}(t) \equiv u(t - T_i) = \begin{cases} 1, t \geq T_i \\ 0, t < T_i \end{cases}, i = 1,2$	To be determined		
$F_{I2}(t) \equiv u(t - T_i) = \begin{cases} 1, t \geq T_i \\ 0, t < T_i \end{cases}, i = 3,4$	To be determined		
$F_{IR1}(t) \equiv u(t - D_1) = \begin{cases} 1, t \geq D_1 \\ 0, t < D_1 \end{cases}$	$D_1 = 1/40$		
$F_{IR2}(t) \equiv u(t - D_2) = \begin{cases} 1, t \geq D_2 \\ 0, t < D_2 \end{cases}$	$D_2 = 1/100$		

Furthermore, due to the fact that TWT inspection and maintenance is costly, a fourth case will be studied, with no minimal maintenance of the TWT but only major maintenance, which should be performed at the last deterioration stage. The results of all cases will be compared.

In addition, these results will be compared with the results of some more cases, which include models with no application of minimal maintenance to the two subsystems and the cases where $F_{pr1} = F_{pr2} = 0$, which means that there is no accelerated deterioration.

Supposing that a military radar unit operates for about 1000 hours per year, availability will be determined by solving Equation 2.4 as follows:

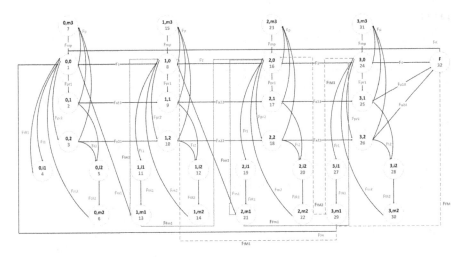

FIGURE 2.2
Case 1 ($g=0, b=2$) model.

$$\max_{T_1,T_2,T_3,T_4} \text{Availability}(T_1,T_2,T_3,T_4) = \max_{T_1,T_2,T_3,T_4} \sum_{i \in A} \pi_i \qquad (2.6)$$

where $0 < T_1 < T_2 < 1000$, $0 < T_3 < T_4 < 1000$.

As an example, a model of case 1 ($g=0$, $b=2$) is presented in Figure 2.2. Imperfect minimal maintenance is depicted with red lines and imperfect major maintenance with dashed red lines.

Furthermore, in Appendices I and II, the calculation of the element $p_{ij}(t)$ of the corresponding matrix and the sojourn times h_i are given as presented by Equations 2.1 and 2.2, respectively.

As derived by Equation 2.4, the availability is given by the following equation:

$$\text{Availability} = \pi_1 + \pi_2 + \pi_3 + \pi_8 + \pi_9 + \pi_{10}$$
$$+ \pi_{16} + \pi_{17} + \pi_{18} + \pi_{24} + \pi_{25} + \pi_{26}$$

The results of the system's maximum availability, as calculated by solving Equation 2.6 for all cases, are presented in Table 2.2. Furthermore, the inspection times for all cases are presented in Table 2.3.

It is clearly shown in Table 2.2 that, comparing the four maintenance policies, a negligible improvement in availability is achieved. Furthermore, there is a negligible difference in availability between the cases with acceleration and without acceleration. This small difference is explained by the following reasons:

- TWTs are constructed with high reliability materials.
- Deterioration rates are small.

TABLE 2.2

Availability Results

		Availability (%)		
Case		Accelerated Deterioration	No Accelerated Deterioration	No Minimal Maintenance of Subsystems
1	$g=0, b=2$	99.25	99.42	98.11
2	$g=0, b=1$	99.24	99.41	98.10
3	$g=1, b=2$	99.32	99.45	98.17
4	No minimal maintenance, $b=2$	99.18	99.31	98.05

- Minimal maintenance that is performed on subsystems corrects accelerated deterioration, thus alleviating its consequences for TWT availability.
- Minimal maintenance that is performed on the TWT alleviates the consequences of accelerated deterioration.
- During sudden failures, maintenance—a correction of malfunctions of subsystems—is performed; thus, the accelerated deterioration is reduced.

The maximum availability is achieved in the case where $g=1$, $b=2$. Very high availability is also achieved in the case where no minimal maintenance is performed on the TWT. This is very interesting, considering that TWT maintenance is too expensive. Moreover, a non-negligible difference is observed in cases where no minimal maintenance is performed on the two subsystems. This difference averages 1.15%, which means that minimal maintenance on the power supply and cooling systems has a non-negligible impact on the transmitter's availability.

From Table 2.3, it can be seen that the best inspection interval for the subsystems is once per year. Concerning TWT, the best inspection interval for cases 3 and 4 is also once per year. For case 2, the appropriate inspection interval is semi-annually, and lastly, for case 1, there is a difference between T_1 and T_2. Specifically, T_1 should be 418 h while T_2 should be 45 h. This case may be uneconomical when one takes into account the high inspection cost of TWT.

TABLE 2.3

Inspection Times

Case		T_1	T_2	T_3	T_4
1	$g=0, b=2$	417.98	44.84	999.99	999.82
2	$g=0, b=1$	521.036	504.42	999.99	999.52
3	$g=1, b=2$	999.908	999.369	999.99	999.891
4	No minimal maintenance, $b=2$	989.145	981.754	999.99	999.894

2.5 Conclusion

In this chapter, a condition-based model was presented for complex systems that undergo accelerated deterioration due to fault interactions. The imperfect maintenance scenario was also taken into account. Judging from the results, the performance of CBM can alleviate the symptoms of accelerated deterioration and improve availability of the system.

The results of this study are not comparable with other studies because of the different approaches to deterioration processes, using the semi-Markov model and simultaneous CBM in major systems and subsystems, and the imperfect maintenance scenario.

An extension to this work could include the studying of maximization of availability and minimization of cost in complex systems, considering more complex and combined failure interactions. Additionally, one could examine how probability distributions such as, for instance, the Weibull distribution might affect the reliability of complex systems that undergo failure interactions. Finally, other modeling methods such as the Markov regenerative process (MRGP) could be studied to compare against the semi-Markov modeling of this chapter.

References

Acton, Q. A., Sulfur hexafluoride, in *Fluorides: Advances in Research and Application*, 2013, Atlanta, Georgia, Scholarly Editions, pp. 113–186.

Advisory Group for Aerospace Research & Development, High power microwaves (HPM), presented at the Sensor and Propagation Panel Symposium held in Ottawa, Canada, 2–5 May 1994, available at: http://www.dtic.mil/dtic/tr/fulltext/u2/a296642.pdf.

Ahmad, R. and S. Kamaruddin, An overview of time-based and condition-based maintenance in industrial application, *Computers & Industrial Engineering*, vol. 63, no. 1, pp. 135–149, Aug. 2012.

Amari, S. V. and L. McLaughlin, Optimal design of a condition-based maintenance model, in *Proc. 2004 Annu. Rel. and Maintainability Symp.* (RAMS '04), pp. 528–533, 2004.

Amari, S. V., L. McLaughlin, and H. Pham, Cost-effective condition based maintenance using Markov decision processes, in *Proc. 2006 Annu. Rel. and Maintainability Symp.* (RAMS '06), pp. 464–469, 2006.

Batun, S. and M. Azizoğlu, Single machine scheduling with preventive maintenances, *International Journal of Production Research*, vol. 47, no. 7, pp. 1753–1771, 2009.

Bazan, J. A. N., Thermal conductivity of poly-alpha-olefin (PAO)-based nanofluids, MS thesis, University of Dayton, Dayton, Ohio, 2010, available at: https://etd.ohiolink.edu/rws_etd/document/get/dayton1282162148/inline.

Bloch, H. P. and F. K. Geitner, The failure analysis and troubleshooting system, in *Practical Machinery Management for Process Plants*, 4th edition, Waltham, MA, Butterworth-Heinemann, 2012, pp. 1–9.

Brown, M. and F. Proschan, Imperfect repair, *Journal of Applied Probability*, vol. 20, no. 4, pp. 851–859, Dec. 1983.

Caballé, N.C., I. T. Castro, C. J. Pérez, and J. M. Lanza-Gutiérrez, A condition-based maintenance of a dependent degradation-threshold-shock model in a system with multiple degradation processes, *Reliability Engineering and System Safety*, vol. 134, pp. 98–109, Feb. 2015.

Cekyay, B. and S. Ozekici, Optimal maintenance of systems with Markovian mission and deterioration, *European Journal of Operational Research*, vol. 219, pp. 123–133, 2012.

Chen, D. and K. S. Trivedi, Analysis of periodic preventive maintenance with general system failure distribution, *Proceedings 2001 Pacific Rim International Symposium on Dependable Computing*, pp. 103–107, 2001.

Chen, D. and K. S. Trivedi, Closed-form analytical results for condition based maintenance, *Reliability Engineering and System Safety*, vol. 76, pp. 43–51, 2002.

Chen, D. and K. S. Trivedi, Optimization for condition-based maintenance with semi-Markov decision process, *Reliability Engineering and System Safety*, vol. 90, pp. 25–29, 2005.

Chiesa, M. and S. K. Das, Experimental investigation of the dielectric and cooling performance of colloidal suspensions in insulating media, *Colloids and Surfaces A: Physicochem. Eng. Aspects*, vol. 335, pp. 88–97, 2009.

Communications and Power Industries, CPI Life Extender Technology, available at: http://www.cpii.com/docs/library/4/LifeExtender%20rev%204.pdf.

Dekker, R., R. E. Wildeman, and F. A. van der Duyn Schouten, A review of multi-component maintenance models with economic dependence, *Mathematical Methods of Operations Research*, vol. 45, no. 3, pp. 411–435, Oct. 1997.

Dhillon, B. S., Corrective maintenance, in *Engineering Maintenance: A Modern Approach*, Florida, CRC Press LLC, 2002, pp. 72–88.

Do, P., A. Voisin, E. Levrat, and B. Iung, A proactive condition-based maintenance strategy with both perfect and imperfect maintenance actions, *Reliability Engineering and System Safety*, vol. 133, pp. 22–32, 2015a.

Do, P., P. Scarf, and B. Iung, Condition-based maintenance for a two-component system with dependencies, *IFAC-Papers On Line*, vol. 48, no. 21, pp. 946–951, Sep. 2015b.

Fei, Y. W., X. L. Peng, L. P. Tong, and H. W. Yang, Study on high temperature oxidative deterioration for poly-α-olefins aviation lubricant materials, *Materials Research Innovations*, vol. 19, Supplement 10, pp. S10-360–S10-365–97, 2015.

Gao, Q. and Y. Ge, Maintenance interval decision models for a system with failure interaction, *Journal of Manufacturing Systems*, vol. 36, pp. 109–114, July 2015.

Golmakani, H. R. and H. Moakedi, Periodic inspection optimization model for a two-component repairable system with failure interaction, *The International Journal of Advanced Manufacturing Technology*, vol. 61, no. 1, pp. 295–302, July 2012.

Khodorkovsky, J., B. Khusid, A. Acrivos, and M. Beltran, Comprehensive electrical evaluation of polyalphaolefin (PAO) dielectric coolant, Final report, prepared for Naval Air Warfare Center – Aircraft Division, November 12 1997, available at: http://www.dtic.mil/dtic/tr/fulltext/u2/a363781.pdf.

Kulkarni, V. G., Markov renewal process, in *Modeling and Analysis of Stochastic Systems*, London, UK, Chapman and Hall, 1995, pp. 477–550.

Leung, K.-N. F. and K.-K. Lai, A preventive maintenance and replacement policy of a series system with failure interaction, *Optimization: A Journal of Mathematical Programming and Operations Research*, vol. 61, no. 2, pp. 223–237, Jan. 2012.

Li, H., E. Deloux, and L. Dieulle, A condition-based maintenance policy for multi-component systems with Lévy copulas dependence, *Reliability Engineering & System Safety*, vol. 149, pp. 44–55, May 2016.

Liang, Z. and A. K. Parlikad, A condition-based maintenance model for assets with accelerated deterioration due to fault propagation, *IEEE Transactions on Reliability*, vol. 64, no. 3, pp. 972–982, Sep. 2015.

Liu, X., J. Li, K. N. Al-Khalifa, A. S. Hamouda, D. W. Coit, and E. A. Elsayed, Condition-based maintenance for continuously monitored degrading systems with multiple failure modes, *IIE Transactions*, vol. 45, no. 4, pp. 422–435, 2013.

Manatos, A., V. P. Koutras, and A. N. Platis, Dependability and performance stochastic modelling of a two-unit repairable production system with preventive maintenance, *International Journal of Production Research*, vol. 54, no. 21, pp. 6395–6415, 2016.

Mudawar, I., Assessment of high-heat-flux thermal management schemes, *IEEE Transactions on Components and Packaging Technologies*, vol. 24, no. 2, pp. 122–141, Jun. 2001

Nelson, I. C., Characterization of thermo-physical properties and forced convective heat tranfer of poly-alpha-olefin (PAO) nanofluids, MS thesis, Texas A&M University, Texas, 2007, available at: http://oaktrust.library.tamu.edu/bitstream/handle/1969.1/ETD-TAMU-1569/NELSON-THESIS.pdf.

Nguyen, K.-A., P. Do, and A. Grall, Multi-level predictive maintenance for multi-component systems, *Reliability Engineering and System Safety*, vol. 144, pp. 83–94, Dec. 2015a.

Rao, P. N. S. and V. N. A. Naikan, An optimization methodology for condition based minimal and major preventive maintenance, *Economic Quality Control*, vol. 21, no. 1, pp. 127–141, 2006.

Rasmekomen, N. and A. Kumar Parlikad, Condition-based maintenance of multi-component systems with degradation state-rate interactions, *Reliability Engineering & System Safety*, vol. 148, pp. 1–10, Apr. 2016.

Rezaei, E., A new model for the optimization of periodic inspection intervals with failure interaction: A case study for a turbine rotor, *Case Studies in Engineering Failure Analysis*, vol. 9, pp. 148–156, Oct. 2017.

Saranga, H., Relevant condition-parameter strategy for an effective condition-based maintenance, *Journal of Quality in Maintenance Engineering*, vol. 8, no. 1, pp. 92–105, 2002.

Shi, A. H. and B. J.-C. Zeng, Real-time prediction of remaining useful life and preventive opportunistic maintenance strategy for multi-component systems considering stochastic dependence, *Computers & Industrial Engineering*, vol. 93, pp. 192–204, Mar. 2016.

Sivan, L., TWT – detailed principles of operation, in *Microwave Tube Transmitters*, 1st edition, London, Chapman and Hall, 2002, pp. 39–123.

Sun, Y., L. Ma, and J. Mathew, Failure analysis of engineering systems with preventive maintenance and failure interactions, *Computers & Industrial Engineering*, vol. 57, pp. 539–549, Aug 2008.

Sung, C.-K., S.-H. Sheu, T.-S. Hsu, and Y.-C. Chen, Extended optimal replacement policy for a two-unit system with failure rate interaction and external shocks, *International Journal of Systems Science*, vol. 44, no. 5, pp. 877–888, 2013.

Tang, D., J. Yu, X. Chen, and V. Makis, An optimal condition-based maintenance policy for a degrading system subject to the competing risks of soft and hard failure, *Computers & Industrial Engineering*, vol. 83, pp. 100–110, May 2015.

Thelander, H., Extending the lifetime of high power TWTAs, *Milsat Magazine*, March 2015, available at: http://www.milsatmagazine.com/story.php?number=1452989667.

Timofeeva, E. V., Nanofluids for heat transfer – potential and engineering strategies, in *Two Phase Flow, Phase Change and Numerical Modeling*, A. Ahsan (Ed.), InTech, 2001, DOI:10.5772/22158, available at: http://www.intechopen.com/books/two-phase-flow-phase-change-and-numerical-modeling/nanofluids-for-heat-transfer-potential-and-engineering-strategies.

Trivedi, K. S., Continuous-time Markov chains, in *Probability and Statistics with Reliability, Queuing and Computer Science Applications*, New York, John Wiley & Sons, 2002, pp. 405–554.

Wang, W., A simulation-based multivariate Bayesian control chart for real time condition-based maintenance of complex systems, *European Journal of Operational Research*, vol. 218, pp. 726–734, May 2012.

Zhang, Z., S. Wu, S. Lee, and J. Ni, Modified iterative aggregation procedure for maintenance optimization of multi-component systems with failure interaction, *International Journal of Systems Science*, vol. 45, no. 12, pp. 2480–2489, 2014.

Zhang, Z., S. Wu, B. Li, and S. Lee, (n, N) type maintenance policy for multi-component systems with failure interactions, *International Journal of Systems Science*, vol. 46, no. 6, pp. 1051–1064, 2015.

Appendix I

According to the analysis in Section 2.2.2, the non-zero elements $p_{ij}(t)$ of the matrix that corresponds to case 1 ($g = 0$, $b = 2$) are calculated using Equation 2.1 as follows:

$$p12 = \int_0^{T_1} f_{pr1} \bar{F}_p \bar{F}_1 \bar{F}_{pr2} (1 - u(t - T_1)) dt = \int_0^{T_1} e^{-\lambda_p t} e^{-\lambda_1 t} e^{-\lambda_{pr2} t} \lambda_{pr1} e^{-\lambda_{pr1} t} \left(1 - \frac{t}{T_1}\right) dt$$

$$p13 = \int_0^{T_1} f_{pr2} \bar{F}_p \bar{F}_1 \bar{F}_{pr1} (1 - u(t - T_1)) dt = \int_0^{T_1} e^{-\lambda_p t} e^{-\lambda_1 t} e^{-\lambda_{pr1} t} \lambda_{pr2} e^{-\lambda_{pr2} t} \left(1 - \frac{t}{T_1}\right) dt$$

$$p17 = \int_0^{T_1} f_p \overline{F}_{pr2} \overline{F}_1 \overline{F}_{pr1} (1 - u(t - T_1)) dt = \int_0^{T_1} \lambda_p e^{-\lambda_p t} e^{-\lambda_1 t} e^{-\lambda_{pr1} t} e^{-\lambda_{pr2} t} \left(1 - \frac{t}{T_1}\right) dt$$

$$p18 = \int_0^{T_1} f_1 \overline{F}_p \overline{F}_{pr2} \overline{F}_{pr1} (1 - u(t - T_1)) dt = \int_0^{T_1} e^{-\lambda_p t} \lambda_1 e^{-\lambda_1 t} e^{-\lambda_{pr1} t} e^{-\lambda_{pr2} t} \left(1 - \frac{t}{T_1}\right) dt$$

$$p14 = 1 - p12 - p13 - p17 - p18$$

$$p27 = \int_0^{T_3} f_p \overline{F}_{a11} (1 - u(t - T_3)) dt = \int_0^{T_3} \lambda_p e^{-\lambda_p t} e^{-\lambda_{a11} t} \left(1 - \frac{t}{T_3}\right) dt$$

$$p29 = \int_0^{T_3} f_{a11} \overline{F}_p (1 - u(t - T_3)) dt = \int_0^{T_3} e^{-\lambda_p t} \lambda_{a11} e^{-\lambda_{a11} t} \left(1 - \frac{t}{T_3}\right) dt$$

$$p25 = 1 - p27 - p29$$

$$p37 = \int_0^{T_3} f_p \overline{F}_{a21} (1 - u(t - T_3)) dt = \int_0^{T_3} \lambda_p e^{-\lambda_p t} e^{-\lambda_{a21} t} \left(1 - \frac{t}{T_3}\right) dt$$

$$p310 = \int_0^{T_3} f_{a21} \overline{F}_p (1 - u(t - T_3)) dt = \int_0^{T_3} e^{-\lambda_p t} \lambda_{a21} e^{-\lambda_{a21} t} \left(1 - \frac{t}{T_3}\right) dt$$

$$p35 = 1 - p37 - p310$$

$$p56 = 1$$

$$p61 = \int_0^\infty f_{m2} dt = \int_0^\infty \lambda_{m2} e^{-\lambda_{m2} t} dt$$

$$p71 = \int_0^\infty f_{mp} dt = \int_0^\infty \lambda_{mp} e^{-\lambda_{mp} t} dt$$

$$p89 = \int_0^{T_1} f_{pr1} \overline{F}_p \overline{F}_2 \overline{F}_{pr2} (1 - u(t - T_1)) dt = \int_0^{T_1} e^{-\lambda_p t} e^{-\lambda_2 t} e^{-\lambda_{pr2} t} \lambda_{pr1} e^{-\lambda_{pr1} t} \left(1 - \frac{t}{T_1}\right) dt$$

$$p810 = \int_0^{T_1} f_{pr2} \overline{F}_p \overline{F}_2 \overline{F}_{pr1} (1 - u(t - T_1)) dt = \int_0^{T_1} e^{-\lambda_p t} e^{-\lambda_2 t} e^{-\lambda_{pr1} t} \lambda_{pr2} e^{-\lambda_{pr2} t} \left(1 - \frac{t}{T_1}\right) dt$$

$$p815 = \int_0^{T_1} f_p \overline{F}_{pr2} \overline{F}_2 \overline{F}_{pr1}(1-u(t-T_1))dt = \int_0^{T_1} e^{-\lambda_{pr2}t} e^{-\lambda_2 t} e^{-\lambda_{pr1}t} \lambda_p e^{-\lambda_p t}\left(1-\frac{t}{T_1}\right)dt$$

$$p816 = \int_0^{T_1} f_2 \overline{F}_{pr2} \overline{F}_p \overline{F}_{pr1}(1-u(t-T_1))dt = \int_0^{T_1} e^{-\lambda_p t} e^{-\lambda_{pr2}t} e^{-\lambda_{pr1}t} \lambda_2 e^{-\lambda_2 t}\left(1-\frac{t}{T_1}\right)dt$$

$$p811 = 1 - p89 - p810 - p815 - p816$$

$$p915 = \int_0^{T_3} f_p \overline{F}_{a12}(1-u(t-T_3))dt = \int_0^{T_3} \lambda_p e^{-\lambda_p t} e^{-\lambda_{a12}t}\left(1-\frac{t}{T_3}\right)dt$$

$$p917 = \int_0^{T_3} f_{a12} \overline{F}_p (1-u(t-T_3))dt = \int_0^{T_3} \lambda_{a12} e^{-\lambda_{a12}p t} e^{-\lambda_p t}\left(1-\frac{t}{T_3}\right)dt$$

$$p912 = 1 - p915 - p917$$

$$p1015 = \int_0^{T_3} f_p \overline{F}_{a22}(1-u(t-T_3))dt = \int_0^{T_3} \lambda_p e^{-\lambda_p t} e^{-\lambda_{a22}t}\left(1-\frac{t}{T_3}\right)dt$$

$$p1018 = \int_0^{T_3} f_{a22} \overline{F}_p (1-u(t-T_3))dt = \int_0^{T_3} e^{-\lambda_p t} \lambda_{a22} e^{-\lambda_{a22}t}\left(1-\frac{t}{T_3}\right)dt$$

$$p1012 = 1 - p1015 - p1018$$

$$p1113 = 1$$

$$p1214 = 1$$

$$p131 = \int_0^\infty f_{m1} \overline{F}_{im1} \overline{F}_{Fm1} dt = \int_0^\infty \lambda_{m1} e^{-\lambda_{m1}t} e^{-\lambda_{im1}t} e^{-\lambda_{Fm1}t} dt$$

$$p138 = \int_0^\infty f_{im1} \overline{F}_{m1} \overline{F}_{Fm1} dt = \int_0^\infty \lambda_{im1} e^{-\lambda_{im1}t} e^{-\lambda_{m1}t} e^{-\lambda_{Fm1}t} dt$$

$$p1316 = \int_0^\infty f_{Fm1} \overline{F}_{im1} \overline{F}_{m1} dt = \int_0^\infty \lambda_{Fm1} e^{-\lambda_{Fm1}t} e^{-\lambda_{im1}t} e^{-\lambda_{m1}t} dt$$

$$p148 = \int_0^\infty f_{m2} dt = \int_0^\infty \lambda_{m2} e^{-\lambda_{m2}t} dt$$

$$p158 = \int_0^\infty f_{mp}dt = \int_0^\infty \lambda_{mp}e^{-\lambda_{mp}t}dt$$

$$p1617 = \int_0^{T_2} f_{pr1}\overline{F}_{pr2}\overline{F}_3\overline{F}_p(1-u(t-T_2))dt = \int_0^{T_2} e^{-\lambda_p t}e^{-\lambda_3 t}e^{-\lambda_{pr2} t}\lambda_{pr1}e^{-\lambda_{pr1} t}\left(1-\frac{t}{T_2}\right)dt$$

$$p1618 = \int_0^{T_2} f_{pr2}\overline{F}_{pr1}\overline{F}_3\overline{F}_p(1-u(t-T_2))dt = \int_0^{T_2} e^{-\lambda_p t}e^{-\lambda_3 t}e^{-\lambda_{pr1} t}\lambda_{pr2}e^{-\lambda_{pr2} t}\left(1-\frac{t}{T_2}\right)dt$$

$$p1623 = \int_0^{T_2} f_p\overline{F}_{pr2}\overline{F}_3\overline{F}_{pr1}(1-u(t-T_2))dt = \int_0^{T_2} e^{-\lambda_{pr1} t}e^{-\lambda_3 t}e^{-\lambda_{pr2} t}\lambda_p e^{-\lambda_p t}\left(1-\frac{t}{T_2}\right)dt$$

$$p1624 = \int_0^{T_2} f_3\overline{F}_{pr2}\overline{F}_{pr1}\overline{F}_p(1-u(t-T_2))dt = \int_0^{T_2} e^{-\lambda_p t}e^{-\lambda_{pr1} t}e^{-\lambda_{pr2} t}\lambda_3 e^{-\lambda_3 t}\left(1-\frac{t}{T_2}\right)dt$$

$$p1619 = 1 - p1617 - p1618 - p1623 - p1624$$

$$p1723 = \int_0^{T_4} f_p\overline{F}_{a13}(1-u(t-T_4))dt = \int_0^{T_4} e^{-\lambda_{a13} t}\lambda_p e^{-\lambda_p t}\left(1-\frac{t}{T_4}\right)dt$$

$$p1725 = \int_0^{T_4} f_{a13}\overline{F}_p(1-u(t-T_4))dt = \int_0^{T_4} e^{-\lambda_p t}\lambda_{a13}e^{-\lambda_{a13} t}\left(1-\frac{t}{T_4}\right)dt$$

$$p1720 = 1 - p1723 - p1725$$

$$p1823 = \int_0^{T_4} f_p\overline{F}_{a23}(1-u(t-T_4))dt = \int_0^{T_4} e^{-\lambda_{a23} t}\lambda_p e^{-\lambda_p t}\left(1-\frac{t}{T_4}\right)dt$$

$$p1826 = \int_0^{T_4} f_{a23}\overline{F}_p(1-u(t-T_4))dt = \int_0^{T_4} e^{-\lambda_p t}\lambda_{a23}e^{-\lambda_{a23} t}\left(1-\frac{t}{T_4}\right)dt$$

$$p1820 = 1 - p1823 - p1826 - p1921$$

$$p2022 = 1$$

$$p218 = \int_0^\infty f_{m1}\overline{F}_{im1}\overline{F}_{Fm1}dt = \int_0^\infty \lambda_{m1}e^{-\lambda_{m1} t}e^{-\lambda_{im1} t}e^{-\lambda_{Fm1} t}dt$$

$$p2116 = \int_0^\infty f_{im1} \overline{F}_{m1} \overline{F}_{Fm1} dt = \int_0^\infty \lambda_{im1} e^{-\lambda_{im1}t} e^{-\lambda_{m1}t} e^{-\lambda_{Fm1}t} dt$$

$$p2124 = \int_0^\infty f_{Fm1} \overline{F}_{im1} \overline{F}_{m1} dt = \int_0^\infty \lambda_{Fm1} e^{-\lambda_{Fm1}t} e^{-\lambda_{im1}t} e^{-\lambda_{m1}t} dt$$

$$p2216 = \int_0^\infty f_{m2} dt = \int_0^\infty \lambda_{m2} e^{-\lambda_{m2}t} dt$$

$$p2316 = \int_0^\infty f_{mp} dt = \int_0^\infty \lambda_{mp} e^{-\lambda_{mp}t} dt$$

$$p2425 = \int_0^{T_2} f_{pr1} \overline{F}_{pr2} \overline{F}_4 \overline{F}_p (1 - u(t - T_2)) dt$$

$$= \int_0^{T_2} e^{-\lambda_p t} e^{-\lambda_4 t} \lambda_{pr1} e^{-\lambda_{pr1}t} e^{-\lambda_{pr2}t} \left(1 - \frac{t}{T_2}\right) dt$$

$$p2426 = \int_0^{T_2} f_{pr2} \overline{F}_{pr1} \overline{F}_4 \overline{F}_p (1 - u(t - T_2)) dt$$

$$= \int_0^{T_2} e^{-\lambda_p t} e^{-\lambda_4 t} \lambda_{pr2} e^{-\lambda_{pr2}t} e^{-\lambda_{pr1}t} \left(1 - \frac{t}{T_2}\right) dt$$

$$p2431 = \int_0^{T_2} f_p \overline{F}_{pr2} \overline{F}_4 \overline{F}_{pr1} (1 - u(t - T_2)) dt$$

$$= \int_0^{T_2} e^{-\lambda_{pr1}t} e^{-\lambda_4 t} \lambda_p e^{-\lambda_p t} e^{-\lambda_{pr2}t} \left(1 - \frac{t}{T_2}\right) dt$$

$$p2432 = \int_0^{T_2} f_4 \overline{F}_{pr2} \overline{F}_{pr1} \overline{F}_p (1 - u(t - T_2)) dt$$

$$= \int_0^{T_2} e^{-\lambda_p t} e^{-\lambda_{pr1}t} \lambda_4 e^{-\lambda_4 t} e^{-\lambda_{pr2}t} \left(1 - \frac{t}{T_2}\right) dt$$

$$p2427 = 1 - p2425 - p2426 - p2431 - p2432$$

$$p2531 = \int_0^{T_4} f_p \overline{F}_{a14}(1-u(t-T_4))dt$$

$$= \int_0^{T4} e^{-\lambda_{a14}t}\lambda_p e^{-\lambda_p t}\left(1-\frac{t}{T_4}\right)dt$$

$$p2532 = \int_0^{T_4} f_{14}\overline{F}_p(1-u(t-T_4))dt$$

$$= \int_0^{T_4} e^{-\lambda_p t}\lambda_{a14}e^{-\lambda_{14}t}\left(1-\frac{t}{T_4}\right)dt$$

$$p2528 = 1 - p2531 - p2532$$

$$p2631 = \int_0^{T_4} f_p \overline{F}_{a24}(1-u(t-T_4))dt$$

$$= \int_0^{T_4} e^{-\lambda_{a24}t}\lambda_p e^{-\lambda_p t}\left(1-\frac{t}{T_4}\right)dt$$

$$p2632 = \int_0^{T_4} f_{24}\overline{F}_p(1-u(t-T_4))dt = \int_0^{T_4} e^{-\lambda_p t}\lambda_{24}e^{-\lambda_{24}t}\left(1-\frac{t}{T_4}\right)dt$$

$$p2628 = 1 - p2631 - p2632 - p2729$$

$$p2830 = 1$$

$$p291 = \int_0^\infty f_M \overline{F}_{IM1}\overline{F}_{IM2}\overline{F}_{IM3}\overline{F}_{FM}dt = \int_0^\infty \lambda_M e^{-\lambda_M t}e^{-\lambda_{IM1}t}e^{-\lambda_{IM2}t}e^{-\lambda_{IM3}t}e^{-\lambda_{FM}t}dt$$

$$p298 = \int_0^\infty f_{IM1}\overline{F}_M \overline{F}_{IM2}\overline{F}_{IM3}\overline{F}_{FM}dt = \int_0^\infty \lambda_{IM1} e^{-\lambda_{IM1}t}e^{-\lambda_M t}e^{-\lambda_{IM2}t}e^{-\lambda_{IM3}t}e^{-\lambda_{FM}t}dt$$

$$p2916 = \int_0^\infty f_{IM2}\overline{F}_{IM1}\overline{F}_M \overline{F}_{IM3}\overline{F}_{FM}dt = \int_0^\infty \lambda_{IM2} e^{-\lambda_{IM2}t}e^{-\lambda_{IM1}t}e^{-\lambda_M t}e^{-\lambda_{IM3}t}e^{-\lambda_{FM}t}dt$$

$$p2924 = \int_0^\infty f_{IM3}\overline{F}_{IM1}\overline{F}_{IM2}\overline{F}_M \overline{F}_{FM}dt = \int_0^\infty \lambda_{IM3} e^{-\lambda_{IM3}t}e^{-\lambda_{IM1}t}e^{-\lambda_{IM2}t}e^{-\lambda_M t}e^{-\lambda_{FM}t}dt$$

Availability Modeling of Complex Systems

$$p2932 = \int_0^\infty f_{FM} \overline{F}_{IM1} \overline{F}_{IM2} \overline{F}_{IM3} \overline{F}_M dt = \int_0^\infty \lambda_{FM} e^{-\lambda_M t} e^{-\lambda_{IM1} t} e^{-\lambda_{IM2} t} e^{-\lambda_{IM3} t} e^{-\lambda_M t} dt$$

$$p3024 = \int_0^\infty f_{m2} dt = \int_0^\infty \lambda_{m2} e^{-\lambda_{m2} t} dt$$

$$p3124 = \int_0^\infty f_{mp} dt = \int_0^\infty \lambda_{mp} e^{-\lambda_{mp} t} dt$$

$$p321 = \int_0^\infty f_R dt = \int_0^\infty \lambda_R e^{-\lambda_R t} dt$$

Appendix II

According to the analysis in Section 2.2.2, the mean sojourn times $h_i(t)$ that correspond to case 1 ($g = 0$, $b = 2$) are calculated using Equation 2.2 as follows:

$$h1 = \int_0^{T_1} \overline{F}_1 \overline{F}_p \overline{F}_{pr1} \overline{F}_{pr2} (1 - u(t - T_1)) dt = \int_0^{T_1} e^{-(\lambda_p + \lambda_1 + \lambda_{pr1} + \lambda_{pr2})t} \left(1 - \frac{t}{T_1}\right) dt$$

$$h2 = \int_0^{T_3} \overline{F}_p \overline{F}_{a11} (1 - u(t - T_3)) dt = \int_0^{T_3} e^{-(\lambda_p + \lambda_{a11})t} \left(1 - \frac{t}{T_3}\right) dt$$

$$h3 = \int_0^{T_3} \overline{F}_p \overline{F}_{a21} (1 - u(t - T_3)) dt = \int_0^{T_3} e^{-(\lambda_p + \lambda_{a21})t} \left(1 - \frac{t}{T_3}\right) dt$$

$h4 = D1$

$h5 = D2$

$$h6 = \int_0^\infty \overline{F}_{m2} dt = \int_0^\infty e^{-\lambda_{m2} t} dt$$

$$h7 = \int_0^\infty \overline{F}_{mp} dt = \int_0^\infty e^{-\lambda_{mp} t} dt$$

$$h8 = \int_0^{T_1} \overline{F}_2 \overline{F}_p \overline{F}_{pr1} \overline{F}_{pr2} (1 - u(t - T_1)) dt = \int_0^{T_1} e^{-(\lambda_p + \lambda_2 + \lambda_{pr1} + \lambda_{pr2})t} \left(1 - \frac{t}{T_1}\right) dt$$

$$h9 = \int_0^{T_3} \overline{F}_p \overline{F}_{a12}(1-u(t-T_3))dt = \int_0^{T_3} e^{-(\lambda_p+\lambda_{a12})t}\left(1-\frac{t}{T_3}\right)dt$$

$$h10 = \int_0^{T_3} \overline{F}_p \overline{F}_{a22}(1-u(t-T_3))dt = \int_0^{T_3} e^{-(\lambda_p+\lambda_{a22})t}\left(1-\frac{t}{T_3}\right)dt$$

$$h11 = D_1$$

$$h12 = D_2$$

$$h13 = \int_0^\infty \overline{F}_{m1}\overline{F}_{im1}\overline{F}_{Fm1}dt = \int_0^\infty e^{-(\lambda_{m1}+\lambda_{im1}+\lambda_{Fm1})t}dt$$

$$h14 = \int_0^\infty \overline{F}_{m2}dt = \int_0^\infty e^{-\lambda_{m2}t}dt$$

$$h15 = \int_0^\infty \overline{F}_{mp}dt = \int_0^\infty e^{-\lambda_{mp}t}dt$$

$$h16 = \int_0^{T_2} \overline{F}_3\overline{F}_p\overline{F}_{pr1}\overline{F}_{pr2}(1-u(t-T_2))dt = \int_0^{T_2} e^{-(\lambda_p+\lambda_3+\lambda_{pr1}+\lambda_{pr2})t}\left(1-\frac{t}{T_2}\right)dt$$

$$h17 = \int_0^{T_4} \overline{F}_p\overline{F}_{a13}(1-u(t-T_4))dt = \int_0^{T_4} e^{-(\lambda_p+\lambda_{a13})t}\left(1-\frac{t}{T_4}\right)dt$$

$$h18 = \int_0^{T_4} \overline{F}_p\overline{F}_{a23}(1-u(t-T_4))dt = \int_0^{T_4} e^{-(\lambda_p+\lambda_{a23})t}\left(1-\frac{t}{T_4}\right)dt$$

$$h19 = D_1$$

$$h20 = D_2$$

$$h21 = \int_0^\infty \overline{F}_{m1}\overline{F}_{im1}\overline{F}_{Fm1}dt = \int_0^\infty e^{-(\lambda_{m1}+\lambda_{im1}+\lambda_{Fm1})t}dt$$

$$h22 = \int_0^\infty \overline{F}_{m2}dt = \int_0^\infty e^{-\lambda_{m2}t}dt$$

$$h23 = \int_0^\infty \overline{F}_{mp}dt = \int_0^\infty e^{-\lambda_{mp}t}dt$$

$$h24 = \int_0^{T_2} \overline{F}_4 \overline{F}_p \overline{F}_{pr1} \overline{F}_{pr2}(1-u(t-T_2))dt = \int_0^{T_2} e^{-(\lambda_p+\lambda_4+\lambda_{pr1}+\lambda_{pr2})t}\left(1-\frac{t}{T_2}\right)dt$$

$$h25 = \int_0^{T_4} \overline{F}_p \overline{F}_{a14}(1-u(t-T_4))dt = \int_0^{T_4} e^{-(\lambda_p+\lambda_{a14})t}\left(1-\frac{t}{T_4}\right)dt$$

$$h26 = \int_0^{T_4} \overline{F}_p \overline{F}_{a24}(1-u(t-T_4))dt = \int_0^{T_4} e^{-(\lambda_p+\lambda_{a24})t}\left(1-\frac{t}{T_4}\right)dt$$

$$h27 = D_1$$

$$h28 = D_2$$

$$h29 = \int_0^{\infty} \overline{F}_M \overline{F}_{IM1} \overline{F}_{IM2} \overline{F}_{IM3} \overline{F}_{FM} dt = \int_0^{\infty} e^{-(\lambda_M+\lambda_{IM1}+\lambda_{IM2}+\lambda_{IM3}\lambda_{FM})t} dt$$

$$h30 = \int_0^{\infty} \overline{F}_{m2} dt = \int_0^{\infty} e^{-\lambda_{m2}t} dt$$

$$h31 = \int_0^{\infty} \overline{F}_{mp} dt = \int_0^{\infty} e^{-\lambda_{mp}t} dt$$

$$h32 = \int_0^{\infty} \overline{F}_R dt = \int_0^{\infty} e^{-\lambda_R t} dt$$

3

Modeling a Multi-State K-out-of-N: G System with Loss of Units

Ruiz-Castro Juan Eloy and Mohammed Dawabsha

CONTENTS

3.1 Introduction 43
3.2 Assumptions and State Space 45
3.3 The Model 46
    3.3.1 Auxiliary Functions 47
    3.3.2 Transition Probability Matrix 47
3.4 Transient Distribution 49
3.5 The Long-Run Distribution 49
3.6 Measures 50
    3.6.1 Availability 50
    3.6.2 Reliability 51
    3.6.3 Conditional Probability of Failure 51
        3.6.3.1 Conditional Probability of r Repairable Failures 51
        3.6.3.2 Conditional Probability of nr Non-Repairable Failures 51
3.7 Markov Counting Process to Calculate the Mean Number of
New Systems 52
3.8 Numerical Example 52
3.9 Conclusions 55
Acknowledgments 55
References 55
Appendix I 56
Appendix II 58

3.1 Introduction

Severe damage and considerable financial losses are caused when poor reliability provokes a system failure. Redundant systems are considered to be of considerable research interest as a means of improving reliability and availability and of avoiding possible catastrophic failure. Various redundant systems have been proposed, including cold standby, warm standby, hot standby, and the k-out-of-n: G system. The latter is an n-system that works if at least

k units are operational. This system, introduced by Birnbaum et al. (1961), is a popular type of redundancy that is applied in various fields, such as electronic, industrial, and military systems. A generalized k-out-of-n system with parallel modules was developed by Cui and Xie (2005). Recently, Kamalja (2017) modeled a generalized k-out-of-n: F system with parallel modules.

Multi-state systems are of particular importance in ensuring reliability. Traditional reliability theory considers systems in which the units perform in terms of binary models composed of up state (performing) and down state (failure). However, many real-life systems contain multiple components with different performance levels. Lisnianski and Frenkel (2012) included Markov processes in the analysis of multi-state systems, highlighting the benefits of their application.

When complex multi-state systems are modeled, intractable expressions may appear and difficulties can arise in the computational implementation and in applicability. It is desirable to achieve expressions in an algorithmic computational form that allow us to readily interpret and apply the results obtained. One class of distributions that makes it possible to model complex systems with well-structured results, thanks to its matrix-algebraic form, is the phase-type distribution (PH), which was introduced and analyzed in detail by Neuts (1975, 1981), who highlighted its useful algorithmic properties. One of the main properties of PH distributions is that they comprise a dense class of distributions within the set of nonnegative probability distributions. Neuts (1975) pointed out that all discrete distributions with finite support can be represented by PH distributions. Multiple redundant systems have been modeled by considering phase-type distributions. Ruiz-Castro and Li (2011) modeled a multi-state k-out-of-n: G system where the embedded lifetimes are PH distributed.

In reliability studies, it is usually assumed that when a system unit undergoes a non-repairable failure it is replaced by a new one within a negligible time. This assumption, however, is not always realistic. Another, perhaps more practical option, is that of redundant systems, in which a unit that undergoes a non-repairable failure will not be replaced while the system is operational. This situation has been analyzed for different redundant multi-state systems (Ruiz-Castro 2018; Ruiz-Castro 2015; Ruiz-Castro and Fernández-Villodre 2012).

In the present study, we model a discrete-time complex multi-state k-out-of-n: G system with loss of units in a well-structured form. The lifetime of the units is governed by PH distributions in which different performance stages are introduced. The units of the system may undergo repairable and/or non-repairable failures. In the first case, the unit goes to the repair facility and in the second, it is removed. The repair time is also PH distributed. Each time that a non-repairable failure occurs, the unit is removed. When fewer than k units remain in the system, it is replaced by a new n-system.

The chapter is organized as follows. In Section 3.2, the system is described in detail, setting out the assumptions made and the state space employed. The model is then described in Section 3.3 in algorithmic form. The transient and the stationary distributions are presented in Sections 3.4 and 3.5, respectively, where matrix-analytic methods are used. Various measures are applied in Section 3.6 and a Markov counting process is developed in Section 3.7 to explain the mean number of new systems obtained. In Section 3.8, a numerical example is given to show the applicability of the model and, finally, in Section 3.9, the main conclusions drawn are presented.

3.2 Assumptions and State Space

We assume a multi-state system composed initially of n independent units subject to different types of failure. Each multi-state unit may experience internal repairable and/or non-repairable failure. In the first case, the unit is transferred to the repair facility, where there is one repairperson. However, if an operational unit undergoes a non-repairable failure, it is removed and not replaced. We thus propose a k-out-of-n: G system, which is operational while at least k units are operational. Therefore, when $n-k+1$ (or more) non-repairable failures occur, the system is replaced by a new, identical one. The system satisfies the following assumptions.

Assumption 1: The lifetime of each unit is discrete-time PH distributed with representation $(\alpha, \mathbf{T})$. The order of the matrix $\mathbf{T}$ is the number of operational stages, m.

Assumption 2: Each unit can undergo a repairable or non-repairable failure. We assume two absorbing states, one for each kind of failure. The probability of failure depends on the operational stage. Thus, the probability of repairable or non-repairable failure is given by the column vectors $\mathbf{T}_r^0$ and $\mathbf{T}_{nr}^0$, respectively. Clearly, the total absorbing vector produced by any transient state is given by $\mathbf{T}^0 = \mathbf{e} - \mathbf{T}\mathbf{e} = \mathbf{T}_r^0 + \mathbf{T}_{nr}^0$.

Assumption 3: The repair time employed by the repairperson is PH distributed with representation $(\beta, \mathbf{S})$ where the order of $\mathbf{S}$ is equal to q (number of repair stages).

Assumption 4: When a non-repairable failure occurs, the unit is removed. The number of units in the system is always greater than or equal to k.

Assumption 5: While the system is operational, each operational unit and each unit under repair can change stage at the same time.

Assumption 6: If the number of units in the system is l, greater than or equal to k, then the system is operational only if at least k units are operational (the number of units under repair is less than or equal to $l-k$). Otherwise, the system is broken, in which case only the repairpersons continue operating, and all other operations cease.

Assumption 7: Unit quality after a repair is as good as new.

Assumption 8: The times involved in the model are independent.

The discrete case is more complex than the continuous one, as several transitions may occur at the same time. The state space of the system, denoted by E, is described as follows.

This state space is composed of macro-states such that $E = \{U^n, U^{n-1}, \ldots, U^k\}$, where U^l denotes the phases when there are l units in the system. In turn, these macro-states are composed of macro-states U_s^l, for $l = k, \ldots, n$ and $s = 0, \ldots, k, l$ units in the system and s units in the repair facility. The phases are given by

$$U_0^l = \{(i_1, \ldots, i_l); 1 \leq i_s \leq m, s = 1, \ldots, l\} \quad \text{for } l = k, \ldots, n$$

$$U_a^l = \{(i_1, \ldots, i_{l-a}; j); 1 \leq i_s \leq m, s = 1, \ldots, l-a; 1 \leq j \leq t\} \quad \text{for } a = 1, \ldots, l-1$$

$$U_l^l = \{j; 1 \leq j \leq t\}$$

where i_s indicates the state of the sth operational unit of the system and j the repair stage.

3.3 The Model

The k-out-of-n system with loss of units is modeled by a vector Markov process with state space as described in the previous section. The transition probability matrix, **P**, is composed of two levels of matrix blocks. The first level has the matrices $\mathbf{R}^{l,h}$ and contains the transition probabilities between the macro-states from U^l to U^h (i.e., from l units in the system to h units in the system), where $h - l$ non-repairable failures occur. The matrix has the following structure:

$$\mathbf{P} = \begin{pmatrix} \mathbf{R}^{n,n} & \mathbf{R}^{n,n-1} & \mathbf{R}^{n,n-2} & \cdots & \mathbf{R}^{n,k+1} & \mathbf{R}^{n,k} \\ \mathbf{R}^{n-1,n} & \mathbf{R}^{n-1,n-1} & \mathbf{R}^{n-1,n-2} & \cdots & \mathbf{R}^{n-1,k+1} & \mathbf{R}^{n-1,k} \\ \mathbf{R}^{n-2,n} & 0 & \mathbf{R}^{n-2,n-2} & \cdots & \mathbf{R}^{n-2,k+1} & \mathbf{R}^{n-2,k} \\ \vdots & \vdots & \ddots & \ddots & \vdots & \vdots \\ \mathbf{R}^{k+1,n} & 0 & 0 & \cdots & \mathbf{R}^{k+1,k+1} & \mathbf{R}^{k+1,k} \\ \mathbf{R}^{k,n} & 0 & 0 & \cdots & 0 & \mathbf{R}^{k,k} \end{pmatrix}.$$

These matrices, $\mathbf{R}^{l,h}$, are composed of matrix blocks (second level). These new matrix blocks, $\mathbf{B}^{l,h}_{i,j}$, contain the transition between the macro-states, from U_i^l (l units in the system of which i are in the repair facility) to U_j^h (h units in the system of which j are in the repair facility).

3.3.1 Auxiliary Functions

To build these matrix blocks, the following auxiliary functions are incorporated, taking into account the phases of the operational units.

1. Transition matrix for the operational units when the system is composed of l units, of which a are in the repair facility, when $l-h$ non-repairable failures and u repairable failures take place, in a specific failure order ($s_1,\ldots,s_u$ is the ordinal of the repairable failures and $k_1,\ldots,k_w$ is the ordinal of the non-repairable failures). This situation is described as $C(l,a,w,u;k_1,\ldots,k_w;s_1,\ldots,s_u)$.
2. Transition matrix for the operational units when the system is composed of l units, of which a are in the repair facility, when w non-repairable failures and u repairable failures take place where the failure order is not established. This situation is described as $b(l,a,w,u)$.
3. If there are l units is the system, a of which are in the repair facility, and the number of non-repairable failures at the next step is greater than or equal to $l-k+1$, then the system has to be replaced. The probability of this occurring during the phases of the system is denoted by $d(k,l,a)$.

These functions are further developed in Appendix I.

3.3.2 Transition Probability Matrix

The transition probability matrix, **P**, is composed of the following block matrices.

For $l = k, \ldots, n,$

$$\mathbf{R}^{l,J} = \begin{pmatrix}
\mathbf{B}_{0,0}^{l,J} & \mathbf{B}_{0,1}^{l,J} & \mathbf{B}_{0,2}^{l,J} & \cdots & \mathbf{B}_{0,J-k-2}^{l,J} & \mathbf{B}_{0,J-k-1}^{l,J} & \mathbf{B}_{0,J-k}^{l,J} & \mathbf{B}_{0,J-k+1}^{l,J} & \cdots & \cdots & \mathbf{B}_{0,J-1}^{l,J} & \mathbf{B}_{0,J}^{l,J} \\
\mathbf{B}_{1,0}^{l,J} & \mathbf{B}_{1,1}^{l,J} & \mathbf{B}_{1,2}^{l,J} & \cdots & \mathbf{B}_{1,J-k-2}^{l,J} & \mathbf{B}_{1,J-k-1}^{l,J} & \mathbf{B}_{1,J-k}^{l,J} & \mathbf{B}_{1,J-k+1}^{l,J} & \cdots & \cdots & \mathbf{B}_{1,J-1}^{l,J} & \mathbf{B}_{1,J}^{l,J} \\
 & \mathbf{B}_{2,1}^{l,J} & \mathbf{B}_{2,2}^{l,J} & \cdots & \mathbf{B}_{2,l-k-2}^{l,J} & \mathbf{B}_{2,J-k-1}^{l,J} & \mathbf{B}_{2,J-k}^{l,J} & \mathbf{B}_{2,J-k+1}^{l,J} & \cdots & \cdots & \mathbf{B}_{2,J-1}^{l,J} & \mathbf{B}_{2,J}^{l,J} \\
 & & \ddots & \ddots & \vdots & \vdots & \vdots & \vdots & \cdots & \cdots & \vdots & \vdots \\
 & & & \ddots & \mathbf{B}_{l-k-1,J-k-2}^{l,J} & \mathbf{B}_{l-k-1,J-k-1}^{l,J} & \mathbf{B}_{l-k-1,J-k}^{l,J} & \mathbf{B}_{l-k-1,J-k+1}^{l,J} & \cdots & \cdots & \mathbf{B}_{l-k-1,J-1}^{l,J} & \mathbf{B}_{l-k-1,J}^{l,J} \\
\mathbf{B}_{l-k,0}^{l,J} & 0 & 0 & \cdots & 0 & \mathbf{B}_{l-k,J-k-1}^{l,J} & \mathbf{B}_{l-k,J-k}^{l,J} & \mathbf{B}_{l-k,J-k+1}^{l,J} & \cdots & \cdots & \mathbf{B}_{l-k,J-1}^{l,J} & \mathbf{B}_{l-k,J}^{l,J} \\
\mathbf{B}_{l-k+1,0}^{l,J} & 0 & 0 & \cdots & \cdots & 0 & \mathbf{B}_{l-k+1,J-k}^{l,J} & \mathbf{B}_{l-k+1,J-k+1}^{l,J} & 0 & \cdots & 0 & 0 \\
\vdots & \vdots & \vdots & & \vdots & & & \ddots & \ddots & & \vdots & \vdots \\
\mathbf{B}_{l-1,0}^{l,J} & 0 & 0 & & 0 & \cdots & \cdots & 0 & \mathbf{B}_{l-1,J-2}^{l,J} & \mathbf{B}_{l-1,J-1}^{l,J} & 0 \\
0 & 0 & 0 & \cdots & \cdots & 0 & \cdots & \cdots & 0 & 0 & \mathbf{B}_{l,J-1}^{l,J} & \mathbf{B}_{l,J}^{l,J}
\end{pmatrix}$$

For $l=k, \ldots, n-1$,

$$\mathbf{R}^{l,n} = \begin{pmatrix} \mathbf{B}_{00}^{l,n} & 0 & 0 & \cdots & 0 & 0 \\ \vdots & 0 & 0 & \cdots & 0 & 0 \\ \mathbf{B}_{\min\{l-k,k-1\},0}^{l,n} & 0 & 0 & \cdots & 0 & 0 \\ 0 & \vdots & \ddots & \ddots & \ddots & \vdots \\ \vdots & \vdots & \ddots & \ddots & \ddots & \vdots \\ 0 & 0 & 0 & \cdots & 0 & 0 \end{pmatrix}_{l \times n}$$

and for $l=k+1, \ldots, n$ and $h=l-1, \ldots, k$,

$$\mathbf{R}^{l,h} = \begin{pmatrix} \mathbf{B}_{00}^{l,h} & \mathbf{B}_{01}^{l,h} & \mathbf{B}_{02}^{l,h} & \cdots & \cdots & \cdots & \mathbf{B}_{0h-2}^{l,h} & \mathbf{B}_{0h-1}^{l,h} & \mathbf{B}_{0h}^{l,h} \\ \mathbf{B}_{10}^{l,h} & \mathbf{B}_{11}^{l,h} & \mathbf{B}_{12}^{l,h} & \cdots & \cdots & \cdots & \mathbf{B}_{1h-2}^{l,h} & \mathbf{B}_{1h-1}^{l,h} & \mathbf{B}_{1h}^{l,h} \\ 0 & \mathbf{B}_{21}^{l,h} & \mathbf{B}_{22}^{l,h} & \cdots & \cdots & \cdots & \mathbf{B}_{2h-2}^{l,h} & \mathbf{B}_{2h-1}^{l,h} & \mathbf{B}_{2h}^{l,h} \\ 0 & 0 & \mathbf{B}_{32}^{l,h} & \cdots & \cdots & \cdots & \mathbf{B}_{3h-2}^{l,h} & \mathbf{B}_{3h-1}^{l,h} & \mathbf{B}_{3h}^{l,h} \\ \vdots & \vdots & \ddots & \ddots & \ddots & & \ddots & \vdots & \vdots \\ 0 & 0 & 0 & \cdots & 0 & \mathbf{B}_{\min\{h,l-k\},\min\{h,l-k\}-1}^{l,h} & \cdots & \mathbf{B}_{\min\{h,l-k\},h-1}^{l,h} & \mathbf{B}_{\min\{h,l-k\},h}^{l,h} \\ 0 & 0 & 0 & \cdots & \cdots & \cdots & 0 & 0 & 0 \\ \vdots & \vdots & \vdots & \ddots & \ddots & \ddots & \vdots & \vdots & \vdots \\ 0 & 0 & 0 & \cdots & \cdots & \cdots & 0 & 0 & 0 \end{pmatrix}_{l+1 \times h+1}$$

The matrix $\mathbf{B}_{ij}^{l,h}$ for $1 \le i \le h-2$ and $i+1 \le j \le h-1$ is described here in detail; the rest of the matrices are given in Appendix II.

Matrix $\mathbf{B}_{ij}^{l,h}$

We assume that there are l units in the system ($l=k+1, \ldots, n$) and that i of them are in the repair facility ($1 \le i \le h-2$). We also assume that the number of the units in the repair facility is less than or equal to $l-k$. Thus, the system is operational and failures may occur. The probability matrix of having h units in the system ($h=l-1, \ldots, k$) with j of them in the repair facility ($i+1 \le j \le h-1$) is given by the following two possibilities.

1. $l-h$ non-repairable and $j-i$ repairable failures occur. The unit in the repair facility is not repaired.

$$b(l,i,l-h,j-i) \otimes \mathbf{S}$$

2. $l-h$ non-repairable and $j-i+1$ repairable failures occur. The unit in the repair facility is repaired and the repair of another one begins.

$$b(l,i,l-h,j-i+1) \otimes \mathbf{S}^0 \boldsymbol{\beta}$$

Thus, the matrix block is given by

$$\mathbf{B}_{ij}^{l,h} = I_{\{i\leq l-k\}}\left\{b(l,i,l-h,j-i)\otimes \mathbf{S} + b(l,i,l-h,j-i+1)\otimes \mathbf{S}^0\boldsymbol{\beta}\right\}$$

3.4 Transient Distribution

The transition probability at m steps can be calculated from the transition probability matrix described above. This is obtained by considering the matrix blocks associated with the macro-states U^l, when there are l units in the system. If the transition probability matrix from macro-state U^i to U^j is denoted by $\mathbf{p}_{U^iU^j}^{(m)}$ for $i, j = k, \ldots, n$ then, in a recursive way, we can obtain

$$\mathbf{p}_{U_iU_j}^{(m)} = \sum_{l=j\cdot I_{\{j\neq n\}}+k\cdot I_{\{j=n\}}}^{n} \mathbf{p}_{U_iU_l}^{(m-1)}\mathbf{R}^{lj}$$

If the system is initially composed of n new units, then the transient distribution is given by

$$\mathbf{p}^{(m)} = (1,0,\ldots,0)\sum_{j=k}^{n}\mathbf{p}_{U^nU^j}^{(m)}$$

$\mathbf{p}_{U^l}^{(m)}$ then denotes the values corresponding to the macro-state derived from $\mathbf{p}^{(m)}$. In an analogous way, $\mathbf{p}_{U_a^l}^{(m)}$ denotes the corresponding value for the macro-state l units in the system, when a of them are in the repair facility.

3.5 The Long-Run Distribution

The long-run distribution, $\boldsymbol{\pi}$, is obtained by matrix-analytic methods. This distribution has been calculated for the macro-states U^l, when there are l units in the system. The stationary probability of being in this macro-state is denoted by $\boldsymbol{\pi}^l$; thus, $\boldsymbol{\pi} = (\boldsymbol{\pi}^n, \boldsymbol{\pi}^{n-1}, \ldots, \boldsymbol{\pi}^k)$. This stationary distribution verifies the balance equations $\boldsymbol{\pi}\mathbf{P} = \boldsymbol{\pi}$ where the stationary distribution verifies the normalization condition. These equations can be expressed as

$$\boldsymbol{\pi}^n = \sum_{s=k}^{n}\boldsymbol{\pi}^s\mathbf{R}^{s,n},$$

$$\pi^l = \sum_{s=l}^{n} \pi^s \mathbf{R}^{s,l}; l = k,\ldots,n-1,$$

$$\sum_{s=k}^{n} \pi^s \mathbf{e} = 1$$

The solution to this system is given by

$$\pi^l = \pi^n \mathbf{R}^l; l = k,\ldots,n-1$$

where

$$\mathbf{R}^l = \left(\mathbf{R}^{n,l} + I_{\{l \le n-2\}} \sum_{s=l+1}^{n-1} \mathbf{R}^s \mathbf{R}^{s,l} \right) \left(\mathbf{I} - \mathbf{R}^{l,l} \right)^{-1}; l = k,\ldots,n-1$$

The vector π^n is obtained from the first balance equation and the normalization condition. This is equal to

$$\pi^n = (1,0) \left[\left(\mathbf{e} + \sum_{s=k}^{n-1} \mathbf{R}^s \mathbf{e} \right) \left(\mathbf{I} - \sum_{s=k}^{n-1} \mathbf{R}^s \mathbf{R}^{s,n} - \mathbf{R}^{n,n} \right)^* \right]^{-1}$$

where the matrix $\mathbf{A}^*$ is a matrix $\mathbf{A}$ without the first column.

π_a^l denotes the corresponding stationary values for the macro-state of l units in the system, of which a are in the repair facility.

3.6 Measures

Several measures associated with the system have been applied, including availability, reliability, and the conditional probability of failure.

3.6.1 Availability

The availability is the probability that at a certain time v the system is operational. In the stationary case, this measure is given by

$$A(v) = \sum_{l=k}^{n} \sum_{a=0}^{l-k} \mathbf{p}_{u_a^l}^{(v)} \mathbf{e}$$

This measure in the stationary case is given by $A = \sum_{l=k}^{n} \sum_{a=0}^{l-k} \pi_a^l \mathbf{e}.$

3.6.2 Reliability

The first time that the system is not operational is a PH distribution, denoted by $((1,0,...,0), \mathbf{P}^*)$, where $\mathbf{P}^*$ is the matrix $\mathbf{P}$ restricted to the macro-state $\bigcup_{l=k}^{n} \bigcup_{a=0}^{l-k} U_a^l$. The reliability function, defined as the probability that at time v the system will be operational before any failure has occurred, is given by

$$R(v) = (1, 0..., 0)(\mathbf{I} - \mathbf{P}^*)^{-1} \mathbf{P}^{*v} \mathbf{P}^{*0}.$$

3.6.3 Conditional Probability of Failure

Two different conditional probabilities of failure are defined; that of repairable failure and that of non-repairable failure. Both are defined in a transient and also in a stationary regime.

3.6.3.1 Conditional Probability of r Repairable Failures

If the system is working at time $v-1$, the probability that at the next time the system will have undergone only r repairable failures is given by

$$CPF_r(v) = \sum_{l=k}^{n} \left[\begin{array}{l} \mathbf{p}_{U_0^l}^{(v-1)} \cdot b(l, k, a=0, r, 0) \cdot \mathbf{e}_{m^l} \\ + \sum_{a=1}^{l-r} \left[\mathbf{p}_{U_a^l}^{(v-1)} \cdot b(l, k, a, r, 0) \otimes \mathbf{e}_t \right] \cdot \mathbf{e}_{t \cdot m^{l-a}} \end{array} \right]$$

3.6.3.2 Conditional Probability of nr Non-Repairable Failures

If the system is working at time $v-1$, the probability that at the next time the system will have undergone only nr non-repairable failures is given by

$$CPF_r(v) = \sum_{l=k}^{n} \left[\begin{array}{l} \mathbf{p}_{U_0^l}^{(v-1)} \cdot b(k, l, a=0, 0, nr) \cdot \mathbf{e}_{m^l} \\ + \sum_{a=1}^{l-r} \left[\mathbf{p}_{U_a^l}^{(v-1)} \cdot b(l, k, a, 0, nr) \otimes \mathbf{e}_t \right] \cdot \mathbf{e}_{t \cdot m^{l-a}} \end{array} \right]$$

3.7 Markov Counting Process to Calculate the Mean Number of New Systems

In this section, a Markov counting process is developed to calculate the mean number of new systems up to a certain time. To achieve this objective, the transition probability matrix from a state to a new system is determined. This matrix is given by

$$\mathbf{D}_{ns} = \begin{pmatrix} \mathbf{D}^{n,n} & 0 & \cdots & 0 \\ \mathbf{D}^{n-1,n} & 0 & \cdots & 0 \\ \vdots & \vdots & \ddots & \vdots \\ \mathbf{D}^{k,n} & 0 & \cdots & 0 \end{pmatrix},$$

where for $l = k, \ldots, n$,

$$\mathbf{D}^{l,n} = \begin{pmatrix} \mathbf{D}^{l,n}_{00} & 0 & 0 & \cdots & 0 & 0 \\ \vdots & 0 & 0 & \cdots & 0 & 0 \\ \mathbf{D}^{l,n}_{\min\{l-k,k-1\},0} & 0 & 0 & \cdots & 0 & 0 \\ 0 & \vdots & \ddots & \ddots & \ddots & \vdots \\ \vdots & \vdots & \ddots & \ddots & \ddots & \vdots \\ 0 & 0 & 0 & \cdots & 0 & 0 \end{pmatrix}_{l \times n}$$

$$\mathbf{D}^{l,n}_{00} = d(k,l,0) \otimes \boldsymbol{\alpha} \otimes \overset{n}{\cdots} \otimes \boldsymbol{\alpha},$$

$i = 1, \ldots, \min\{l-k, k-1\}$; $\mathbf{D}^{l,n}_{i0} = d(k,l,i) \otimes \boldsymbol{\alpha} \otimes \overset{n}{\cdots} \otimes \boldsymbol{\alpha} \otimes \mathbf{e}$.

Then, the mean number of new systems up to a certain v is given by

$$\phi(v) = (1,0,\ldots,0) \sum_{i=0}^{v-1} \mathbf{P}^i \mathbf{D} \mathbf{e} \tag{3.1}$$

In the stationary case it is given by $\phi = \pi \mathbf{D} \mathbf{e}$, the number of new systems per unit of time.

3.8 Numerical Example

Let us assume a 2-out-of-4: G system where the time of each operational unit is PH distributed with representation $(\boldsymbol{\alpha}, \mathbf{T})$ and where

$$\boldsymbol{\alpha} = (1,0,0), \mathbf{T} = \begin{pmatrix} 0.98 & 0.01 & 0.002 \\ 0 & 0.98 & 0.01 \\ 0 & 0 & 0.99 \end{pmatrix}$$

Each operational unit may undergo a repairable or a non-repairable failure. These failures can occur from states 1, 2, or 3, respectively, in accordance with the following column probability vectors:

$$\mathbf{T}_r^0 = \begin{pmatrix} 0.005 \\ 0.0075 \\ 0.0075 \end{pmatrix}, \mathbf{T}_{nr}^0 = \begin{pmatrix} 0.003 \\ 0.0025 \\ 0.0025 \end{pmatrix}$$

The mean time until a failure occurs is 110 units of time (u.t.), with mean time to repairable failure of 80 u.t. and mean time to non-repairable failure of 30 u.t.

When a repairable failure occurs, the unit is transferred to the repair facility. The repair time is PH distributed with representation ($\boldsymbol{\beta}$, $\mathbf{S}$) where

$$\boldsymbol{\beta} = (1,0), \mathbf{S} = \begin{pmatrix} 0.2 & 0.4 \\ 0.1 & 0.5 \end{pmatrix}$$

The mean repair time is equal to 2.5 u.t. Various measures can be applied. Thus, in a stationary regime, the ratio in each macro-state is given by the stationary distribution, as shown in Table 3.1.

The availability function is shown in Figure 3.1.

The stationary availability is equal to 0.9846. This value is the operational time ratio and can be derived from Table 3.1. The probability of the system being operational at a certain time before the first system failure is given in Figure 3.2.

The mean time up to the first failure of the system is equal to 346.0609 u.t. Finally, Table 3.2 shows the mean number of new systems calculated for different times from Equation 3.1.

TABLE 3.1

Proportional Time in Each Macro-State

Number of Units in the Repair Facility	Number of Units in the System		
	4	3	2
0	0.2109	0.2913	0.4554
1	0.0125	0.0141	0.0149
2	0.0005	0.0004	0.0000
3	0.0000	0.0000	
4	0.0000		

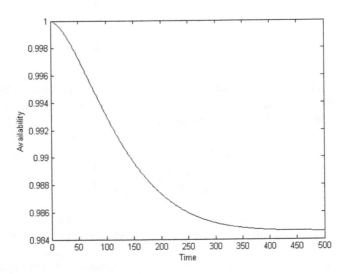

FIGURE 3.1
Availability of the system.

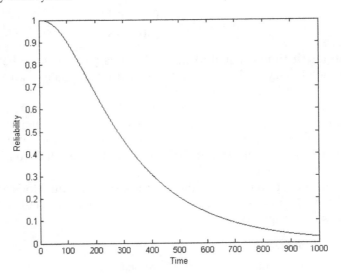

FIGURE 3.2
Reliability function.

TABLE 3.2

Mean Number of New Systems up to a Certain Time

	Mean Number of New Systems									
v	100	200	300	400	500	600	700	800	900	1000
$\phi(t)v$	0.0468	0.2071	0.4286	0.6702	0.9164	1.1629	1.4091	1.6552	1.9012	2.1472

The stationary value, proportional to the number of new systems per unit of time, is 0.0025.

3.9 Conclusions

This chapter describes a multi-state complex k-out-of-n: G system, modeled in an algorithmic and computational form. The system is operational when at least k units are operational. Both repairable and non-repairable failures are included in the system. When a non-repairable failure occurs, the unit is removed. When the number of units in the system is less than or equal to $n-k-1$, it is considered to be nonoperational and is replaced by a new one. At present, the system is modeled only for the case of internal failures, but more complex systems could also be modeled by the same approach.

The transient distribution and the long-run case have also been computed, and several interesting reliability measures obtained, for both transient and stationary regimes. The mean number of new systems per unit of time is calculated by means of a Markov counting process. A numerical example is given to show the versatility of the model.

Acknowledgments

This chapter is partially supported by the Junta de Andalucía, Spain, under grant FQM-307, by the Ministerio de Economía y Competitividad, Spain, under grant MTM2013-47929-P, and by the European Regional Development Fund (ERDF).

References

Birnbaum, Z.W., Esary, J.D., Saunders, S.C. (1961) Multicomponent systems and structures and their reliability. *Technometrics*, 3, 55–77.

Cui, L. and Xie, M. (2005) On a generalized k-out-of-n system and its reliability. *International Journal of Systems Science*, 36, 5, 267–274.

Kamalja, K.K. (2017) Reliability computing method for a generalized k-out-of-n system. *Journal of Computational and Applied Mathematics*, 323, 111–122.

Lisnianski, A. and Frenkel, I. (2012) *Recent advances in system reliability: Signatures, multi-state systems and statistical inference*. London: Springer-Verlag.

Neuts, M.F. (1981) *Matrix Geometric Solutions in Stochastic Models. An Algorithmic Approach*. Baltimore: Johns Hopkins University Press.

Neuts, M.F. (1975) Probability Distributions of Phase Type, *Liber amicorum professor emeritus H. Florin*. Belgium: Department of Mathematics, University of Louvain, 183–206.

Ruiz-Castro, J.E. (2018) A D-MMAP to model a complex multi-state system with loss of units. In: *Recent Advances in Multi-state Systems Reliability*. A. Lisnianski, I. Frenkel, and A. Karagrigoriou (Eds), pp. 39–59. Springer International Publishing, AG 2018, Switzerland.

Ruiz-Castro, J.E. (2015) A preventive maintenance policy for a standby system subject to internal failures and external shocks with loss of units. *International Journal of Systems Science*, 46, 9, 1600–1613.

Ruiz-Castro, J.E. and Fernández-Villodre, G. (2012) A complex discrete warm standby system with loss of units. *European Journal of Operational Research*, 218, 456–469.

Ruiz-Castro, J.E. and Li, Q.L. (2011) Algorithm for a general discrete k-out-of-n: G system subject to several types of failure with an indefinite number of repair-persons. *European Journal of Operational Research*, 211, 97–111.

Appendix I

1. Transition matrix for the operational units when the system is composed of l units, a of them in the repair facility, $l-h$ non-repairable failures occur and u repairable failures take place when the failure order is determined ($s_1,\ldots,s_u$ the ordinal of the repairable failures and $k_1,\ldots,k_w$ the ordinal for the non-repairable failures).

For $h < l$ and $u > 0$,

$$C(l,a,w,u;k_1,\ldots,k_w;s_1,\ldots,s_u) = \mathbf{T}(1) \otimes \cdots \otimes \mathbf{T}(l-a)$$

where

$$\mathbf{T}(v) = \begin{cases} \mathbf{T}_r^0; & v = s_1,\ldots,s_u \\ \mathbf{T}_{nr}^0; & v = k_1,\ldots,k_w \\ \mathbf{T}; & \text{otherwise} \end{cases}$$

This function is denoted as follows for the following cases:

- For $w=0$ and $u>0$, $C(l,a,w=0,u;s_1,\ldots,s_u) = \mathbf{T}(1) \otimes \cdots \otimes \mathbf{T}(l-a)$
- For $u=0$ and $w>0$, $C(l,a,w,u=0;k_1,\ldots,k_{l-h}) = \mathbf{T}(1) \otimes \cdots \otimes \mathbf{T}(l-a)$
- For $w=0$ and $u=0$, $C(l,a,w=0,u=0) = \overset{l-a}{\mathbf{T} \otimes \cdots \otimes \mathbf{T}}$
- For $w=l-a$, $C(l,a,w=l-a,u=0) = \overset{l-a}{\mathbf{T}_{nr}^0 \otimes \cdots \otimes \mathbf{T}_{nr}^0}$

- For $u=l-a$, $C(l,a,w=0,u=l-a) = \mathbf{T}_r^0 \overset{l-a}{\otimes \cdots \otimes} \mathbf{T}_r^0$

2. Transition matrix for the operational units when the system is composed of l units, a of them are in the repair facility, w non-repairable failures occur and u repairable failures have place where the failure order is not determined.

   For $w>0$ and $u>0$,

$$b(l,a,w,u) = \sum_{k_1=1}^{l-a-w+1}\sum_{k_2=k_1+1}^{l-a-w+2}\cdots\sum_{k_w=k_{w-1}+1}^{l-a}\sum_{\substack{s_1=1\\s_1\neq k_v\\v=1,\ldots,w}}^{l-a-u+1}\sum_{\substack{s_2=s_1+1\\s_2\neq k_v\\v=1,\ldots,w}}^{l-a-u+2}\cdots\sum_{\substack{s_u=s_{u-1}+1\\s_u\neq k_v\\v=1,\ldots,w}}^{l-a} C(l,a,w,u;k_1,\ldots,k_w;s_1,\ldots,s_u)$$

According to the specific cases described, we have the following notation:

- For $w=0$ and $l-a\neq u>0$,

$$b(l,a,w=0,u) = \sum_{s_1=1}^{l-a-u+1}\sum_{s_2=s_1+1}^{l-a-u+2}\cdots\sum_{s_u=s_{u-1}+1}^{l-a} C(l,a,0,u;s_1,\ldots,s_u)$$

- For $u=0$ and $l-a\neq w>0$,

$$b(l,a,w,u=0) = \sum_{k_1=1}^{l-a-w+1}\sum_{k_2=k_1+1}^{l-a-w+2}\cdots\sum_{k_w=k_{w-1}+1}^{l-a} C(l,a,w,0;k_1,\ldots,k_w)$$

- For $w=0$ and $u=0$, $b(l,a,w=0,u=0) = c(l,a,0,0) = \mathbf{T} \overset{l-a}{\otimes \cdots \otimes} \mathbf{T}$ and if $a=l$, $b(l,l,w=0,u=0)=1$

- For $w=l-a$, $b(l,a,w=l-a,u=0) = C(l,a,l-a,0) = \mathbf{T}_{nr}^0 \overset{l-a}{\otimes \cdots \otimes} \mathbf{T}_{nr}^0$

- For $u=l-a$, $b(l,a,w=0,u=l-a) = C(l,a,0,l-a) = \mathbf{T}_r^0 \overset{l-a}{\otimes \cdots \otimes} \mathbf{T}_r^0$

3. If there are l units is the system, a of them in the repair facility, and the number of non-repairable failures is greater or equal than $l-k+1$, then the system has to be replaced. The probability that this occurs, by considering the phases of the system, is

$$d(k,l,a) = \sum_{w=l-k+1}^{l-a} d(l,a,w)$$

where

$$d(l,a,w) = \sum_{k_1=1}^{l-a-w+1}\sum_{k_2=k_1+1}^{l-a-w+2}\cdots\sum_{k_w=k_{w-1}+1}^{l-a} D(l,a,w;k_1,\ldots,k_w)$$

and

$$D(l,a,w;k_1,\ldots,k_w) = \mathbf{T}(1)\otimes\cdots\otimes\mathbf{T}(l-a) \text{ with}$$

$$\mathbf{T}(v) = \begin{cases} \mathbf{T}_{nr}^0 & ; \quad v=k_1,\ldots,k_w \\ \mathbf{e}-\mathbf{T}_{nr}^0 & ; \quad \text{otherwise.} \end{cases}$$

Appendix II

The transition probability matrix, **P**, is composed by the following block matrices.

For $l = k, \ldots, n,$

$$R^{l,l} = \begin{pmatrix}
\mathbf{B}_{00}^{l,l} & \mathbf{B}_{01}^{l,l} & \mathbf{B}_{02}^{l,l} & \cdots & \mathbf{B}_{0,l-k-2}^{l,l} & \mathbf{B}_{0,l-k-1}^{l,l} & \mathbf{B}_{0,l-k}^{l,l} & \mathbf{B}_{0,l-k+1}^{l,l} & \cdots & \cdots & \mathbf{B}_{0,l-1}^{l,l} & \mathbf{B}_{0,l}^{l,l} \\
\mathbf{B}_{1,0}^{l,l} & \mathbf{B}_{1,1}^{l,l} & \mathbf{B}_{1,2}^{l,l} & \cdots & \mathbf{B}_{1,l-k-2}^{l,l} & \mathbf{B}_{1,l-k-1}^{l,l} & \mathbf{B}_{1,l-k}^{l,l} & \mathbf{B}_{1,l-k+1}^{l,l} & \cdots & \cdots & \mathbf{B}_{1,l-1}^{l,l} & \mathbf{B}_{1,l}^{l,l} \\
 & \mathbf{B}_{2,1}^{l,l} & \mathbf{B}_{2,2}^{l,l} & \cdots & \mathbf{B}_{2,l-k-2}^{l,l} & \mathbf{B}_{2,l-k-1}^{l,l} & \mathbf{B}_{2,l-k}^{l,l} & \mathbf{B}_{2,l-k+1}^{l,l} & \cdots & & \mathbf{B}_{2,l-1}^{l,l} & \mathbf{B}_{2,l}^{l,l} \\
 & & \ddots & \ddots & & & & & & & \vdots & \vdots \\
 & & & \ddots & \mathbf{B}_{l-k-1,l-k-2}^{l,l} & \mathbf{B}_{l-k-1,l-k-1}^{l,l} & \mathbf{B}_{l-k-1,l-k}^{l,l} & \mathbf{B}_{l-k-1,l-k+1}^{l,l} & \cdots & \cdots & \mathbf{B}_{l-k-1,l-1}^{l,l} & \mathbf{B}_{l-k-1,l}^{l,l} \\
\mathbf{B}_{l-k,0}^{l,l} & 0 & 0 & \cdots & 0 & \mathbf{B}_{l-k,l-k-1}^{l,l} & \mathbf{B}_{l-k,l-k}^{l,l} & \mathbf{B}_{l-k,l-k+1}^{l,l} & \cdots & \cdots & \mathbf{B}_{l-k,l-1}^{l,l} & \mathbf{B}_{l-k,l}^{l,l} \\
\mathbf{B}_{l-k+1,0}^{l,l} & 0 & 0 & \cdots & \cdots & 0 & \mathbf{B}_{l-k+1,l-k}^{l,l} & \mathbf{B}_{l-k+1,l-k+1}^{l,l} & 0 & \cdots & 0 & 0 \\
\vdots & \vdots & \vdots & & & \vdots & & & \ddots & & \vdots & \vdots \\
\mathbf{B}_{l-1,0}^{l,l} & 0 & 0 & & & 0 & \cdots & \cdots & 0 & \mathbf{B}_{l-1,l-2}^{l,l} & \mathbf{B}_{l-1,l-1}^{l,l} & 0 \\
0 & 0 & 0 & \cdots & \cdots & 0 & \cdots & \cdots & 0 & 0 & \mathbf{B}_{l,l-1}^{l,l} & \mathbf{B}_{l,l}^{l,l}
\end{pmatrix}$$

$$\mathbf{B}_{00}^{l,l} = b(l,0,0,0) + I_{\{l=n\}}d(k,l,0)\otimes\overset{n}{\alpha\otimes\cdots\otimes\alpha}$$

$$1\leq j\leq l; \mathbf{B}_{0j}^{l,l} = b(l,0,0,j)\otimes\beta$$

$$\mathbf{B}_{10}^{l,l} = I_{\{1\leq l-k\}}\left[b(l,a=1,0,0)\otimes\alpha\otimes\mathbf{S}^0 + I_{\{l=n,a\leq k-1\}}d(k,l,1)\otimes\overset{n}{\alpha\otimes\cdots\otimes\alpha}\otimes\mathbf{e}\right]$$

$$+ I_{\{1>l-k\}}\left[\mathbf{I}\otimes\overset{l-1}{\ldots}\otimes\mathbf{I}\otimes\alpha\otimes\mathbf{S}^0\right]$$

$$2\leq i\leq n-1; \mathbf{B}_{i0}^{l,l} = I_{\{i\leq l-k;l=n;i\leq k-1\}}d(k,l,i)\otimes\overset{n}{\alpha\otimes\cdots\otimes\alpha}\otimes\mathbf{e}$$

$$1\leq i\leq l-1; \mathbf{B}_{ii}^{l,l} = I_{\{i\leq l-k\}}\left[b(l,i,0,0)\otimes\mathbf{S} + b(l,i,0,1)\otimes\alpha\otimes\mathbf{S}^0\beta\right] + I_{\{i>l-k\}}\mathbf{I}\otimes\overset{l-i}{\ldots}\otimes\mathbf{I}\otimes\mathbf{S}$$

$$\mathbf{B}_{ll}^{l,l} = \mathbf{S}$$

$$2\leq i\leq l; \mathbf{B}_{i,i-1}^{l,l} = \left[I_{\{i\leq l-k\}}b(l,i,0,0) + I_{\{i>l-k\}}\left\{I_{\{i\neq l\}}\left(\mathbf{I}\otimes\overset{l-i}{\ldots}\otimes\mathbf{I}\right) + I_{\{i=l\}}\right\}\right]\otimes\alpha\otimes\mathbf{S}^0\beta$$

$1 \le i \le l-2; i+1 \le j \le l-1; \mathbf{B}_{ij}^{l,l} = I_{\{i \le l-k\}}\left[b(l,i,0,j-i) \otimes \mathbf{S} + b(l,i,0,j-i+1) \otimes \boldsymbol{\alpha} \otimes \mathbf{S}^0\boldsymbol{\beta}\right]$

$1 \le i \le l-1; \mathbf{B}_{i,l}^{l,l} = I_{\{i \le l-k\}}b(l,i,0,l-i) \otimes \mathbf{S}$

For $l = k, \ldots, n-1$,

$$\mathbf{R}^{l,n} = \begin{pmatrix} \mathbf{B}_{00}^{l,n} & 0 & 0 & \cdots & 0 & 0 \\ \vdots & 0 & 0 & \cdots & 0 & 0 \\ \mathbf{B}_{\min\{l-k,k-1\},0}^{l,n} & 0 & 0 & \cdots & 0 & 0 \\ 0 & \vdots & \ddots & \ddots & \ddots & \vdots \\ \vdots & \vdots & \ddots & \ddots & \ddots & \vdots \\ 0 & 0 & 0 & \cdots & 0 & 0 \end{pmatrix}_{l \times n}$$

$\mathbf{B}_{00}^{l,n} = d(k,l,0) \otimes \boldsymbol{\alpha} \otimes \overset{n}{\cdots} \otimes \boldsymbol{\alpha}$

$i = 1,\ldots,l-1; \mathbf{B}_{i0}^{l,n} = I_{\{i \le k-1\}}d(k,l,i) \otimes \boldsymbol{\alpha} \otimes \overset{n}{\cdots} \otimes \boldsymbol{\alpha} \otimes \mathbf{e}$

For $l = k+1, \ldots, n$ and $h = l-1, \ldots, k$,

$$\mathbf{R}^{l,h} = \begin{pmatrix} \mathbf{B}_{00}^{l,h} & \mathbf{B}_{01}^{l,h} & \mathbf{B}_{02}^{l,h} & \cdots & \cdots & \cdots & \mathbf{B}_{0h-2}^{l,h} & \mathbf{B}_{0h-1}^{l,h} & \mathbf{B}_{0h}^{l,h} \\ \mathbf{B}_{10}^{l,h} & \mathbf{B}_{11}^{l,h} & \mathbf{B}_{12}^{l,h} & \cdots & \cdots & \cdots & \mathbf{B}_{1h-2}^{l,h} & \mathbf{B}_{1h-1}^{l,h} & \mathbf{B}_{1h}^{l,h} \\ 0 & \mathbf{B}_{21}^{l,h} & \mathbf{B}_{22}^{l,h} & \cdots & \cdots & \cdots & \mathbf{B}_{2h-2}^{l,h} & \mathbf{B}_{2h-1}^{l,h} & \mathbf{B}_{2h}^{l,h} \\ 0 & 0 & \mathbf{B}_{32}^{l,h} & \cdots & \cdots & \cdots & \mathbf{B}_{3h-2}^{l,h} & \mathbf{B}_{3h-1}^{l,h} & \mathbf{B}_{3h}^{l,h} \\ \vdots & \vdots & \ddots & \ddots & \ddots & \ddots & \ddots & \vdots & \vdots \\ 0 & 0 & 0 & \cdots & 0 & \mathbf{B}_{\min\{h,l-k\},\min\{h,l-k\}-1}^{l,h} & \cdots & \mathbf{B}_{\min\{h,l-k\},h-1}^{l,h} & \mathbf{B}_{\min\{h,l-k\},h}^{l,h} \\ 0 & 0 & 0 & \cdots & \cdots & \ddots & 0 & 0 & 0 \\ \vdots & \vdots & \vdots & \ddots & \ddots & \ddots & \vdots & \vdots & \vdots \\ 0 & 0 & 0 & \cdots & \cdots & \cdots & 0 & 0 & 0 \end{pmatrix}_{l+1 \times h+1}$$

$\mathbf{B}_{00}^{l,h} = b(l,0,l-h,0)$

$1 \le i \le h-1; \mathbf{B}_{ii}^{l,h} = b(l,i,l-h,0) \otimes \mathbf{S} + b(l,i,l-h,1) \otimes \mathbf{S}^0\boldsymbol{\beta}$

$\mathbf{B}_{hh}^{l,h} = b(l,h,l-h,0) \otimes \mathbf{S}$

$\mathbf{B}_{10}^{l,h} = b(l,1,l-h,0) \otimes \mathbf{S}^0$

$2 \le i \le h; \mathbf{B}_{i,i-1}^{l,h} = b(l,i,l-h,0) \otimes \mathbf{S}^0\boldsymbol{\beta}$

$1 \le j \le h; \mathbf{B}_{0j}^{l,h} = b(l,0,l-h,j) \otimes \boldsymbol{\beta}$

$1 \le i \le h-2; i+1 \le j \le h-1; \mathbf{B}_{ij}^{l,h} = b(l,i,l-h,j-i) \otimes \mathbf{S} + b(l,i,l-h,j-i+1) \otimes \mathbf{S}^0\boldsymbol{\beta}$

$1 \le i \le h-1; \mathbf{B}_{i,h}^{l,h} = b(l,i,l-h,h-i) \otimes \mathbf{S}$

4

Reliability Analysis of Multi-State Cloud-RAID with Imperfect Element-Level Coverage

Lavanya Mandava, Liudong Xing,
Vinod M. Vokkarane, and Ola Tannous

CONTENTS

4.1 Introduction 61
4.2 Cloud-RAID 5 and Reliability Modeling 63
4.3 Preliminary Models 64
    4.3.1 Element-Level Coverage for MSS 65
    4.3.2 Multi-State Multi-Valued Decision Diagram (MMDD) 66
4.4 Proposed Evaluation Methods 67
    4.4.1 Disk-Level State Probability Evaluation 67
    4.4.2 System-Level State Probability Evaluation 69
        4.4.2.1 Good State Probability of Cloud-RAID 5 70
        4.4.2.2 Covered Failure State Probability of Cloud-RAID 5 72
        4.4.2.3 Degraded State Probability of Cloud-RAID 5 73
        4.4.2.4 Uncovered Failure State Probability of Cloud-RAID 5..... 74
4.5 Numerical Results and Analysis 75
    4.5.1 Disk-Level Evaluation 75
    4.5.2 System-Level Evaluation 77
        4.5.2.1 Perfect Controller 77
        4.5.2.2 Controller with Perfect Coverage 77
        4.5.2.3 Controller with Imperfect Coverage 78
4.6 Conclusion and Future Work 79
References 80

4.1 Introduction

A cloud storage system can be defined as a network of remote servers used to store, manage, and process data over the Internet (Erl et al., 2013; Deng et al., 2010). Users should be able to access or retrieve their data anytime, anywhere. Any interruption or failure in the service could cause negative effects on the reputation and business of the cloud service providers. Therefore, it is crucial to enhance data reliability in the cloud.

Different solutions have been proposed for enhancing reliability of cloud storage systems. For example, erasure coding was implemented to achieve reliable cloud storage in Lu and Xia (2015). A content storage and delivery scheme was suggested to tolerate failures of cloud servers in Zhang et al. (2013a). A game-theoretic mechanism based on replications (Li et al., 2012) was developed to obtain reliable strategies for users and cloud storage providers in Lin and Tzeng (2014). A novel storage scheme called *assembling chain of erasure coding and replication (ASSER)* was proposed in Yin et al. (2017) for reliable input/output (I/O) performance of cloud storage systems at a much lower storage cost. In Li et al. (2016), a cost-effective cloud data reliability management mechanism was modeled using a proactive replica checking approach. A diamond topology was introduced in Luo et al. (2017) to improve performance and reliability of cloud storage systems.

As another effective solution, cloud-RAIDs (redundant arrays of independent disks) utilize various data redundancy management techniques (Bausch, 2014; Jin et al., 2009) to achieve high data reliability (Fitch and Xu, 2013; Zhang et al., 2013b; Liu and Xing, 2015b). Some research efforts have been dedicated to the reliability modeling and analysis of the cloud-RAID systems at different levels. For example, two-level hierarchical methods combining continuous-time Markov chains and combinatorial multi-valued decision diagrams were developed in Liu and Xing (2015a,b) to assess reliability of cloud-RAID 5 and 6 storage systems with heterogeneous disks from different providers. However, these methods fail to consider an inherent behavior of any fault-tolerant system using an automatic fault recovery mechanism. Such a mechanism, which is responsible for fault detection/location/isolation and system reconfiguration, is seldom fully reliable in practice. When the mechanism malfunctions, the undetected or uncovered fault may propagate, causing extensive damages to the entire system. This behavior is known as *imperfect fault coverage (IPC)* (Myers, 2010). Failure to address this behavior can lead to inaccurate (often overestimated) system reliability results, thus misleading the system design, operation, and optimization activities (Xiang et al., 2015; Li and Mao, 2016). Therefore, it is important to consider IPC in the cloud-RAID system reliability modeling and analysis.

In recent works (Mandava and Xing, 2017; Mandava et al., 2016), effects of IPC were addressed in the reliability analysis of binary-state cloud storage systems, where the system and each disk are assumed to be either operational or failed. In the real-world system, however, the system and disk may exhibit multiple performance levels or states, ranging from perfect operation to complete failure. In other words, a cloud-RAID system can be a multi-state system (MSS). Therefore, the existing binary-state reliability models developed for cloud-RAID systems become inadequate. In this chapter, we extend the state of the art by proposing multi-state reliability modeling of cloud-RAID systems subject to IPC. The cloud-RAID 5 system is used as the case study to illustrate the proposed methodology.

The remainder of the chapter is organized as follows. In Section 4.2, the architecture of the example cloud-RAID 5 system and its reliability modeling using multi-state fault trees are described. Section 4.3 presents preliminary models used in the proposed approach. In Section 4.4, methods for evaluating the disk-level and system-level state probabilities of the example cloud-RAID 5 system are presented. In Section 4.5, numerical results of the example system are presented. Section 4.6 gives conclusions and directions of future research.

4.2 Cloud-RAID 5 and Reliability Modeling

In cloud-RAID systems, data are separated into blocks for storage. These blocks are striped across multiple independent disks (possibly from different cloud service providers) forming an array (Patterson et al., 1989). Particularly in cloud-RAID 5, single-bit parity code is implemented to achieve fault tolerance. Blocks on the same row of the array form a virtual disk drive, with one of them storing the parity information. The parity stripes are also distributed across the disks. Figure 4.1 illustrates an example of cloud-RAID 5 with five disks. This system can tolerate failure of any single disk drive; data stripes on the failed disk can be restored using remaining data stripes and the parity stripes. Therefore, under the binary-state model, the example cloud-RAID 5 system is considered to be reliable if at least four disks are operating correctly.

In cloud-RAID systems, the physical disk drives are often functionally dependent on a RAID controller (Gilroy and Irvine, 2006) that controls the

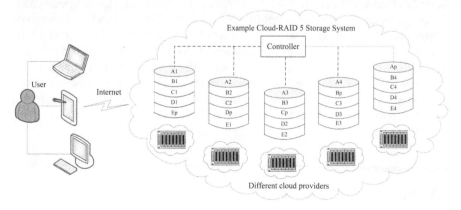

FIGURE 4.1
Architecture of an example cloud-RAID 5 system.

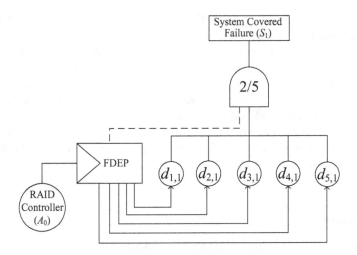

FIGURE 4.2
MFT of the example system being in state CF.

disks so that they work as a logical unit. The controller is assumed to assume binary states: failure (denoted by A_0) and operation (A_1).

Each disk drive and the entire cloud-RAID system can assume four disjoint states: good (G), degraded (D), covered failure (CF), and uncovered failure (UF) represented by 3, 2, 1, and 0, respectively. These states are explained in detail in Section 4.4. As an illustration here, the entire system is in the CF state when two out of five disks are in CF states ($d_{i,1}$ where $i=1, 2, 3, 4, 5$) while the remaining disks occupy either good ($d_{i,3}$) or degraded ($d_{i,2}$) states. Ignoring the UF state ($d_{i,0}$) (as any disk being in a UF state leads to failure of the entire system), Figure 4.2 shows the multi-state fault tree (MFT) model of the example cloud-RAID 5 system being in the CF state, S_1. The functional dependence (FDEP) gate is used to model the five disk drives, being functionally dependent on the RAID controller (Junges et al., 2016). Specifically, the occurrence of a trigger event (the RAID controller) makes the dependent components (the five disk drives) inaccessible or unusable. Though the failure of the controller does not directly lead to system CF, the CF state of disk drives due to controller failure contribute to the CF state of the entire system.

Figure 4.3 illustrates the MFT model of the example system being in the good state, S_3, where four out of the five disks must be in the good state ($d_{i,3}$) and the controller must be operational.

4.3 Preliminary Models

This section describes the IPC model and the multi-state multi-valued decision diagram (MMDD) used in the proposed methodology.

Reliability Analysis of Multi-State Cloud-RAID

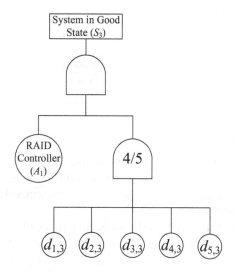

FIGURE 4.3
MFT of the example system being in state G.

4.3.1 Element-Level Coverage for MSS

There exist different models to describe IPC behavior (Shrestha et al., 2010; Levitin and Amari, 2008, 2009). When the effectiveness of the system recovery mechanism relies on the occurrence of individual element faults, the element-level coverage (ELC) model is applicable or appropriate (Myers and Rauzy, 2008). Under ELC, the fault coverage probability of an element does not depend on the states of other system elements. Also, a system with multiple elements can tolerate multiple co-existing single element faults depending on the available redundancy.

Figure 4.4 illustrates the ELC model for a binary-state element (Amari et al., 1999). The entry point denotes the occurrence of an element fault, and the three exits denote possible outcomes. The R exit is reached when the fault is transient and can be handled without discarding any element. The

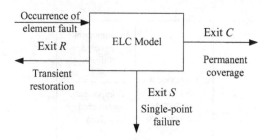

FIGURE 4.4
ELC model for binary-state component. (Adapted from Amari, S. V. et al., *IEEE Trans. Reliab.* 48, 267–274, 1999.)

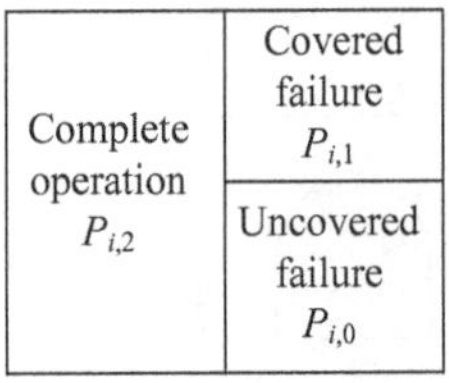

	Covered failure $P_{i,1}$
Complete operation $P_{i,2}$	
	Uncovered failure $P_{i,0}$

FIGURE 4.5
Event space of component under ELC.

C exit is reached when the fault is permanent, and the faulty element has to be discarded. The S exit is reached when the element fault (by itself) leads to failure of the entire system. The three exits are respectively characterized by probabilities (r, c, and s) that sum to 1.

Figure 4.5 presents the event and probability space of element i. The binary states are extended to three states, that is, complete operation, CF, and UF, whose probabilities are represented as $P_{i,2}$, $P_{i,1}$, and $P_{i,0}$, respectively.

In Chang et al. (2005), the ELC model was extended for multi-state elements. Figure 4.6 illustrates the event space for element i with $m+1$ states, where the element being in the UF and CF states is represented as state 0 and state 1 with probabilities $P_{i,0}$ and $P_{i,1}$, respectively. State m is the perfect operation state, with probability $P_{i,m}$, and the remaining states represent various degraded performance levels, respectively, with probabilities $P_{i,m-1}, \ldots, P_{i,2}$.

4.3.2 Multi-State Multi-Valued Decision Diagram (MMDD)

MMDD is a rooted, directed acyclic graph used to represent multi-valued logic functions (Shrestha et al., 2007; Akers et al., 2008). MMDD consists of two sink nodes, 0 and 1, representing the system not being or being in a particular state, respectively. Figure 4.7 (left-hand subfigure) shows the structure of an MMDD model where the non-sink node x represents an n-state component, modeling a multi-valued expression f represented by a *case* format as

$$f = x_1 f_{x=1} + x_2 f_{x=2} + \ldots + x_n f_{x=n} = \text{case}(x, f_1, f_2, \ldots f_n) \tag{4.1}$$

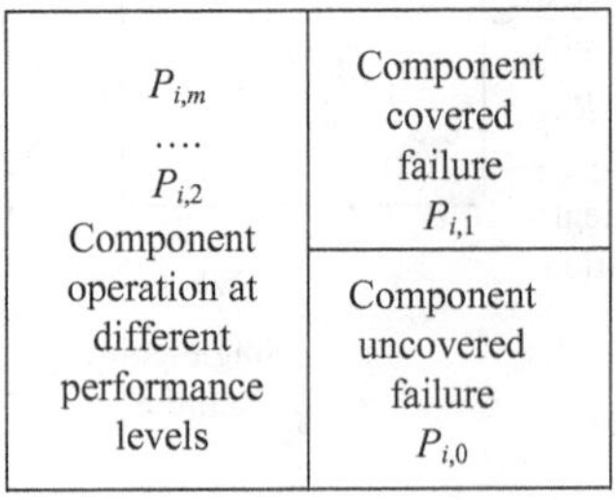

$P_{i,m}$ .... $P_{i,2}$ Component operation at different performance levels	Component covered failure $P_{i,1}$
	Component uncovered failure $P_{i,0}$

FIGURE 4.6
Event space of multi-state component.

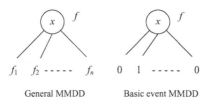

FIGURE 4.7
MMDD model.

The right-hand subfigure in Figure 4.7 shows an MMDD modeling component x being in a specific state 2. The manipulation rules for generating the system MMDD model are given by Equation 4.2, where g and h represent two sub-MMDD models, $\Diamond$ represents the logic operation (AND or OR) for combining the two sub-MMDD models, and *index* represents the ordering of the variable (Xing and Dai, 2009; Xing and Amari, 2015).

$$g \Diamond h = \text{case}(x, G_1, \ldots, G_n) \Diamond \text{case}(y, H_1, \ldots, H_n)$$

$$= \begin{cases} \text{case}(x, G_1 \Diamond H_1, \ldots, G_n \Diamond H_n) & \text{index}(x) = \text{index}(y) \\ \text{case}(x, G_1 \Diamond h, \ldots, G_n \Diamond h) & \text{index}(x) < \text{index}(y) \\ \text{case}(y, g \Diamond H_1, \ldots, g \Diamond H_n) & \text{index}(x) > \text{index}(y) \end{cases} \quad (4.2)$$

With the system MMDD model for a particular state S_k, the probability of the system being in S_k can be evaluated by adding the probabilities of all paths from root node to sink node 1.

4.4 Proposed Evaluation Methods

This section presents methods for evaluating state probabilities of each disk and the entire cloud-RAID system.

4.4.1 Disk-Level State Probability Evaluation

Each disk drive in RAID without considering ELC can assume three states: good (G), degraded (D), and failed (F) states (Jin et al., 2011; Pham et al., 1996). Figure 4.8 illustrates the possible transitions among the three states, which are characterized by the corresponding transition rates. The self-recovery mechanism can make the disk transit from state D to state G with a recovery rate of μ_{dg}. However, if the recovery mechanism fails, the disk enters the permanent failure state F with a rate of λ_{df}.

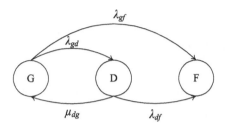

FIGURE 4.8
State transition without IPC. (Adapted from Liu, Q and Xing, L., *Proceedings of International Conference on Quality, Reliability, Risk, Maintenance, and Safety Engineering*. Beijing, China, pp. 1–7, 2015.)

Note that the coverage factors associated with the transitions may be dependent not only on the starting state but also the destination state of the transitions. Transitions originating from the same state but having different destination states imply that they can be caused by different types of faults, which may correspond to different fault detection mechanisms and thus different fault detection probabilities.

Considering IPC, each disk drive in RAID can assume four states: G, D, CF, and UF, denoted by 3, 2, 1, and 0, respectively. Figure 4.9 shows the state transition diagram of a single disk considering the ELC model of Figure 4.4. The single transition due to a fault in Figure 4.8 is extended to three transitions corresponding to the three exits with probabilities r, c, and s, respectively. For example, the disk transits from state G to state D with transition rate of λ_{gd}.

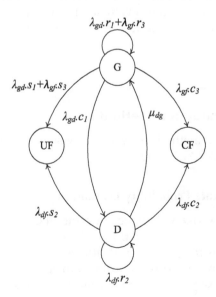

FIGURE 4.9
State transition considering ELC.

c_1 if the fault is covered. If the fault is transient, state G has a self-loop with transition rate $\lambda_{gd}.r_1$; if single-point failure occurs, the disk enters state UF with transition rate of $\lambda_{gd}.s_1$.

Based on the Markov model of Figure 4.9, the time-dependent state probabilities of each disk can be obtained by evaluating Equation 4.3.

$$\begin{bmatrix} 0 & 0 & \lambda_{gd}s_2 & \lambda_{gd}s_1 + \lambda_{gf}s_3 \\ 0 & 0 & \lambda_{df}c_2 & \lambda_{gf}c_3 \\ 0 & 0 & -(\lambda_{gd}s_2 + \lambda_{df}c_2 + \mu_{dg}) & \lambda_{gd}c_1 \\ 0 & 0 & \mu_{dg} & -(\lambda_{gd}s_1 + \lambda_{gf}s_3 + \lambda_{gf}c_3 + \lambda_{gd}c_1) \end{bmatrix} \cdot \begin{bmatrix} P_0(t) \\ P_1(t) \\ P_2(t) \\ P_3(t) \end{bmatrix} = \begin{bmatrix} P_0'(t) \\ P_1'(t) \\ P_2'(t) \\ P_3'(t) \end{bmatrix} \quad (4.3)$$

Applying the Laplace transform-based method, we have the Laplace-transformed state probabilities P^*_3 (s), P^*_2 (s), P^*_1 (s), and P^*_0 (s) as follows:

$$P^*_3(s) = \frac{-(\mu_{dg} + \lambda_{df}s_2 + \lambda_{df}c_2 + s)}{(\mu_{dg}\lambda_{gd}c_1 - \mu_{dg}s + a\mu_{dg} - \lambda_{df}s_2s + \lambda_{df}s_2a - \lambda_{df}c_2s + \lambda_{df}c_2a - s^2 + as)} \quad (4.4)$$

where $a = -(\lambda_{gd}s_1 + \lambda_{gf}s_3 + \lambda_{gf}c_3 + \lambda_{gd}c_1)$

Letting P^*_3 (s) = x, we have

$$P^*_2(s) = \frac{sx - ax - 1}{\mu_{dg}} \quad (4.5)$$

$$P^*_1(s) = \frac{\lambda_{gf}c_3 x}{s} + \frac{\lambda_{df}c_2(sx - ax - 1)}{\mu_{dg}s} \quad (4.6)$$

$$P^*_0(s) = \frac{(\lambda_{gd}s_1 + \lambda_{gf}s_3)x}{s} + \frac{\lambda_{df}s_2(sx - ax - 1)}{\mu_{dg}s} \quad (4.7)$$

Disk state probabilities $P_3(t)$, $P_2(t)$, $P_1(t)$, and $P_0(t)$ can be obtained by finding the inverse Laplace transforms of $P^*_3(s)$, $P^*_2(s)$, $P^*_1(s)$, and $P^*_0(s)$, respectively.

4.4.2 System-Level State Probability Evaluation

As defined in Section 4.2, the example cloud-RAID 5 system is in state G if four out of five disks are in the good state. If two out of five disks are in the CF state, then the system is in the CF state. Any state between the states G and CF is state D. Any single disk of the cloud-RAID 5 system being in the UF state leads to the system being in the UF state. The system states G, D, CF, and UF are represented by S_3, S_2, S_1, and S_0, respectively.

Using the total probability law, the probability of the example cloud-RAID 5 system being in state $k=0, 1, 2, 3$ can be given as (Xing et al., 2014)

$$P_{S_k} = \Pr(S_k \mid A \text{ has UF}) * \Pr(A \text{ has UF})$$
$$+ \Pr(S_k \mid A \text{ has no UF}) * \Pr(A \text{ has no UF}) \quad (4.8)$$
$$= \Pr(S_k \mid A \text{ has UF}) * (1 - P_{u-A}) + P_k * P_{u-A}$$

where:
P_{u-A} is the probability that controller A experiences no UF and is evaluated as $P_{u-A} = (1 - P_{A,0})$
P_k represents the system state probability given that controller A has no UF

Again, by using the total probability law, P_k is obtained by Equation 4.9:

$$P_k = \Pr(S_k \mid A \text{ has CF}) * \Pr(A \text{ has CF})$$
$$+ \Pr(S_k \mid A \text{ has no CF}) * \Pr(A \text{ has no CF}) \quad (4.9)$$
$$= \Pr(S_k \mid \bar{A}) * \Pr(\bar{A}) + \Pr(S_k \mid A) * \Pr(A)$$

where:
$\Pr(\bar{A}) = P_{A,1} / (1 - P_{A,0})$
$\Pr(A) = 1 - \Pr(\bar{A})$

4.4.2.1 Good State Probability of Cloud-RAID 5

From Equation 4.8, the probability of system being in state G may be given by Equation 4.10:

$$P_{S_3} = \Pr(S_3 \mid A \text{ has UF}) * \Pr(A \text{ has UF})$$
$$+ \Pr(S_3 \mid A \text{ has no UF}) * \Pr(A \text{ has no UF}) \quad (4.10)$$
$$= 0 * (1 - P_{u-A}) + P_3 * P_{u-A}$$

where, according to Equation 4.9, P_3 can be evaluated using Equation 4.11. Note that $\Pr(S_3 \mid \bar{A}) = 0$ because when the controller fails covered, the system cannot occupy state G.

$$P_3 = \Pr(S_3 \mid A \text{ has CF}) * \Pr(A \text{ has CF})$$
$$+ \Pr(S_3 \mid A \text{ has no CF}) * \Pr(A \text{ has no CF}) \quad (4.11)$$
$$= \Pr(S_3 \mid \bar{A}) * \Pr(\bar{A}) + \Pr(S_3 \mid A) * \Pr(A) = \Pr(S_3 \mid A) * \Pr(A)$$

Reliability Analysis of Multi-State Cloud-RAID

Applying the total probability law again, $\Pr(S_3|A)$ in Equation 4.11 is calculated by Equation 4.12, where effects of UF of each disk i are separated from the solution combinatorics.

$$\Pr(S_3 | A) = \Pr(S_3 | i \text{ has UF}) * \Pr(i \text{ has UF})$$
$$+ \Pr(S_3 | i \text{ has no UF}) * \Pr(i \text{ has no UF}) \quad (4.12)$$
$$= 0 * (1 - P_{u-i}) + P^c_{S_3} * P_{u-i}$$

$P^c_{S_3}$ in Equation 4.12 can be evaluated using the MMDD model reviewed in Section 4.3.2. Particularly, ternary decision diagrams (TDDs) are used as three disk states (G, D, CF) for each disk i involved in the evaluation. Based on the definition of system state G (four out of five disks being in state G leads to system state G), Figure 4.10 shows the TDD model of the example cloud-RAID 5 system being in state G. Sink node 1 (0) represents the system being (not being) in state G.

From TDDs, $P^c_{S_3}$ can be obtained by adding the probabilities of all paths from the root node to sink node 1, as shown in Equation 4.13. Note that $P^c_{i,3}$ represents the conditional probability of each disk i being in state 3, given that it experiences no UF.

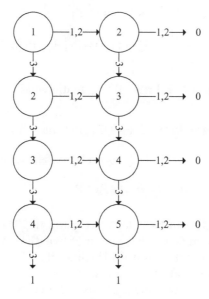

FIGURE 4.10
TDD of system being in state G.

$$P^c_{S_3} = P^c_{1,3}P^c_{2,3}P^c_{3,3}P^c_{4,3} + (1-P^c_{1,3})P^c_{2,3}P^c_{3,3}P^c_{4,3}P^c_{5,3} + P^c_{1,3}(1-P^c_{2,3})P^c_{3,3}P^c_{4,3}P^c_{5,3}$$
$$+ P^c_{1,3}P^c_{2,3}(1-P^c_{3,3})P^c_{4,3}P^c_{5,3} + P^c_{1,3}P^c_{2,3}P^c_{3,3}(1-P^c_{4,3})P^c_{5,3} \qquad (4.13)$$

$$\text{where } P^c_{i,3} = \frac{P_{i,3}}{(1-P_{i,0})}$$

4.4.2.2 Covered Failure State Probability of Cloud-RAID 5

According to Equation 4.8, the probability of the example cloud-RAID 5 system being in state CF (P_{S1}) can be given as

$$P_{S_1} = \Pr(S_1 \mid A \text{ has UF}) * \Pr(A \text{ has UF})$$
$$+ \Pr(S_1 \mid A \text{ has no UF}) * \Pr(A \text{ has no UF}) \qquad (4.14)$$
$$= 0 * (1 - P_{u-A}) + P_1 * P_{u-A}$$

According to Equation 4.9, P_1 in Equation 4.14 can be evaluated by Equation 4.15, where $\Pr(S_1 \mid \bar{A}) = 1$ (the system is in the CF state when A fails covered).

$$P_1 = \Pr(S_1 \mid A \text{ has CF}) * \Pr(A \text{ has CF})$$
$$+ \Pr(S_1 \mid A \text{ has no CF}) * \Pr(A \text{ has no CF}) \qquad (4.15)$$
$$= \Pr(S_1 \mid \bar{A}) * \Pr(\bar{A}) + \Pr(S_1 \mid A) * \Pr(A) = \Pr(\bar{A}) + \Pr(S_1 \mid A) * \Pr(A)$$

Similarly to Equation 4.12, $\Pr(S_1 \mid A)$ in Equation 4.15 can be evaluated as

$$\Pr(S_1 \mid A) = \Pr(S_1 \mid i \text{ has UF}) * \Pr(i \text{ has UF})$$
$$+ \Pr(S_1 \mid i \text{ has no UF}) * \Pr(i \text{ has no UF}) \qquad (4.16)$$
$$= 0 * (1 - P_{u-i}) + P^c_{S_1} * P_{u-i}$$

$P^c_{S_1}$ in Equation 4.16 can be calculated using the TDD model. Based on the definition of system state CF (two out of five disks being in the CF state leads to the system CF state), Figure 4.11 shows the TDD model of the system being in state CF. Sink node 1 (0) represents the system being (not being) in state CF. The TDD evaluation gives $P^c_{S_1}$ as in Equation 4.17, where the terms within square brackets represent the probabilities of all paths leading to sink node 0.

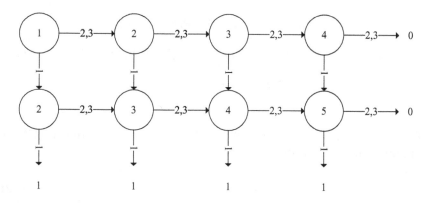

FIGURE 4.11
TDD of system being in state CF.

$$P^c_{S_1} = 1 - \big[Q^c_{1,1}Q^c_{2,1}Q^c_{3,1}Q^c_{4,1} + (1-Q^c_{1,1})Q^c_{2,1}Q^c_{3,1}Q^c_{4,1}Q^c_{5,1}$$
$$+ Q^c_{1,1}(1-Q^c_{2,1})Q^c_{3,1}Q^c_{4,1}Q^c_{5,1} + Q^c_{1,1}Q^c_{2,1}(1-Q^c_{3,1})Q^c_{4,1}Q^c_{5,1}$$
$$+ Q^c_{1,1}Q^c_{2,1}Q^c_{3,1}(1-Q^c_{4,1})Q^c_{5,1} \big] \quad (4.17)$$

$$\text{where} \quad Q^c_{i,1} = 1 - P^c_{i,1} \text{ and } P^c_{i,1} = \frac{P_{i,1}}{(1-P_{i,0})}$$

4.4.2.3 Degraded State Probability of Cloud-RAID 5

Similar to the derivation of probabilities of states G and CF, according to Equations 4.8 and 4.9, the probability of the system being in state D can be derived as follows:

$$P_{S_2} = \Pr(S_2 \mid A \text{ has UF}) * \Pr(A \text{ has UF})$$
$$+ \Pr(S_2 \mid A \text{ has no UF}) * \Pr(A \text{ has no UF}) \quad (4.18)$$
$$= 0 * (1 - P_{u-A}) + P_2 * P_{u-A}$$

where

$$P_2 = \Pr(S_2 \mid A \text{ has CF}) * \Pr(A \text{ has CF})$$
$$+ \Pr(S_2 \mid A \text{ has no CF}) * \Pr(A \text{ has no CF})$$
$$= \Pr(S_2 \mid \overline{A}) * \Pr(\overline{A}) + \Pr(S_2 \mid A) * \Pr(A) \quad (4.19)$$
$$= \Pr(S_2 \mid A) * \Pr(A)$$

$$\Pr(S_2 \mid A) = \Pr(S_2 \mid i \text{ has UF}) * \Pr(i \text{ has UF})$$
$$+ \Pr(S_2 \mid i \text{ has no UF}) * \Pr(i \text{ has no UF}) \quad (4.20)$$
$$= 0 * (1 - P_{u-i}) + P_{S_2}^c * P_{u-i}$$

Equation 4.21 gives $P_{S_2}^c$ in Equation 4.20 as, given that there is no UF, the remaining three state probabilities sum to 1.

$$P_{S_2}^c = 1 - P_{S_3}^c - P_{S_1}^c \quad (4.21)$$

4.4.2.4 Uncovered Failure State Probability of Cloud-RAID 5

According to Equation 4.8, the UF state probability of the example cloud-RAID 5 system is given by Equation 4.22.

$$P_{S_0} = \Pr(S_0 \mid A \text{ has UF}) * \Pr(A \text{ has UF})$$
$$+ \Pr(S_0 \mid A \text{ has no UF}) * \Pr(A \text{ has no UF}) \quad (4.22)$$
$$= 1 * (1 - P_{u-A}) + P_0 * P_{u-A}$$

According to Equation 4.9, P_0 is evaluated in Equation 4.23 where the probability of the system being in the UF state when A is failed covered is 0, that is, $\Pr(S_0 \mid \overline{A}) = 0$.

$$P_0 = \Pr(S_0 \mid A \text{ has CF}) * \Pr(A \text{ has CF})$$
$$+ \Pr(S_0 \mid A \text{ has no CF}) * \Pr(A \text{ has no CF})$$
$$= \Pr(S_0 \mid \overline{A}) * \Pr(\overline{A}) + \Pr(S_0 \mid A) * \Pr(A) \quad (4.23)$$
$$= \Pr(S_0 \mid A) * \Pr(A)$$

$\Pr(S_0 \mid A)$ in Equation 4.23 can be directly obtained by using the probability of any disk being in state UF represented by $P_{i,0}$, as shown in Equation 4.24.

$$\Pr(S_0 \mid A) = 1 - (1 - P_{i,0})^5 \quad (4.24)$$

4.5 Numerical Results and Analysis

4.5.1 Disk-Level Evaluation

Calculating inverse Laplace transforms for Equations 4.4 through 4.7, the disk state probabilities can be obtained. Let us assume that the transition and recovery rates of all five disks are identical, with $\lambda_{gd}=0.01/h$, $\lambda_{gf}=0.001/h$, $\lambda_{df}=0.005/h$, $\mu_{dg}=0.5/h$, $c_1=c_2=c_3=c$, $r_1=r_2=r_3=r$, $s_1=s_2=s_3=s$, $t=1000h$. Table 4.1 lists different sets of disk coverage factors, for which the disk state probabilities are calculated and shown in Table 4.2. Under set 1, all the faults are transient and can be recovered without discarding any disks, so the probability of the disk being in state G is 1. Set 2 corresponds to the perfect fault coverage case, so the probability of the disk being in state UF is 0. Set 3 corresponds to another special case where all faults are a single point of failure. In this case, the disk has the highest UF. Given the same r (set 4 and set 6), for the set with the greater value of c, the probabilities of the disk being in states G, D, and CF are higher.

Table 4.3 shows the results of disk state probabilities for different time values under set 6. Figure 4.12 shows the results graphically. As time proceeds, the probability of the disk being in state G or D declines, while the probability of it being in state CF or UF increases.

We also study effects of different disk transition rates from state G to D on the disk state probabilities. Table 4.4 presents the disk state probabilities for

TABLE 4.1

Disk Coverage Factors

Set	c	r	s
1	0	1	0
2	1	0	0
3	0	0	1
4	0.095	0.9	0.005
5	0.197	0.8	0.003
6	0.099	0.9	0.001

TABLE 4.2

Disk State Probabilities (for c, r, s Sets in Table 4.1)

Set	$P_{i,3}$	$P_{i,2}$	$P_{i,1}$	$P_{i,0}$
1	1	0	0	0
2	0.333761	0.006623	0.659616	0
3	0.000017	3.26E-25	0	0.999983
4	0.858508	0.00163	0.088857	0.051005
5	0.789041	0.003104	0.178481	0.029375
6	0.893378	0.001768	0.094453	0.010401

TABLE 4.3

Disk State Probabilities (for Different Time Values)

t (h)	$P_{i,3}$	$P_{i,2}$	$P_{i,1}$	$P_{i,0}$
1,000	0.893378	0.001768	0.094453	0.010401
2,000	0.799703	0.001582	0.179003	0.019712
5,000	0.573599	0.001135	0.383081	0.042186
10,000	0.329666	0.000652	0.603251	0.066431

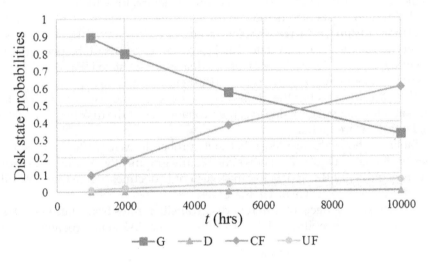

FIGURE 4.12
Disk state probabilities.

different λ_{gd} using parameters of $\lambda_{gf}=0.001$, $\lambda_{df}=0.005$, $\mu_{dg}=0.5$, $c_1=c_2=c_3=0.099$, $r_1=r_2=r_3=0.9$, $s_1=s_2=s_3=0.001$, $t=1000$ h. From Table 4.4, it can be seen that as λ_{gd} increases, the disk good state probability decreases and the degraded state probability increases, which is intuitive. The first set shows a special case where $\lambda_{gd}=\lambda_{gf}=0.001$. Though the transition rates of good to degraded and good to failure are equal, the disk CF and UF state probabilities are both higher than the disk degraded state probability due to the degraded to failure transition and the recovery transition.

TABLE 4.4

Disk State Probabilities (for Different G to D State Transition Rates)

λ_{gd}	$P_{i,3}$	$P_{i,2}$	$P_{i,1}$	$P_{i,0}$
0.001	0.903683	0.000179	0.094235	0.001903
0.005	0.899084	0.000889	0.094332	0.005694
0.01	0.893378	0.001768	0.094453	0.010401
0.02	0.882105	0.00349	0.094689	0.019715

4.5.2 System-Level Evaluation

In this section, we present results of system state probabilities under three cases: the controller A is perfectly reliable, has perfect fault coverage, and has imperfect fault coverage. Note that though the system-level analysis in this section is performed for homogeneous disks, the proposed MMDD-based combinatorial method is also applicable to heterogeneous disks with different disk state probabilities.

4.5.2.1 Perfect Controller

Applying the system state probability evaluation method of Section 4.4.2, Table 4.5 shows the system state probabilities of the example cloud-RAID 5 system when controller A is perfectly reliable. Disk state probabilities under sets 4, 5, and 6 of Table 4.2 (general cases) are used. Results in Table 4.5 show that the system state probabilities depend on the disk state probabilities; the higher the probability of the disk being in a particular state, the higher the probability of the system being in the same state. For example, the system good state probability increases as the disk good state probability increases (sets 5, 4, 6 in increasing order). The system UF state probability increases as the disk UF state probability increases (sets 6, 5, 4 in increasing order).

4.5.2.2 Controller with Perfect Coverage

Table 4.6 shows the cloud-RAID 5 system state probabilities when controller A has no UF ($s_A = 0$) for different combinations of r_A and c_A. The disk state probabilities of set 6 in Table 4.2, controller failure rate $\lambda_A = 0.0001/h$, and

TABLE 4.5

System State Probabilities (A Is Perfectly Reliable)

Set	G	D	CF	UF
4	0.712134	0.001856	0.055709	0.230302
5	0.657764	0.005523	0.198219	0.138494
6	0.875549	0.002411	0.071105	0.050934

TABLE 4.6

System State Probabilities (A Has No UF)

c_A	r_A	G	D	CF	UF
0	1	0.875549	0.002411	0.071105	0.050934
1	0	0.792229	0.002182	0.159501	0.046087
0.7	0.3	0.817225	0.002251	0.132983	0.047541
0.5	0.5	0.833889	0.002297	0.115303	0.048511
0.3	0.7	0.850553	0.002343	0.097624	0.049480

mission time $t_A = 1000$ h are used in the evaluation. The first set of Table 4.6 ($r_A = 1$ means that the controller faults are always transient) gives a special case where the system state probabilities are identical to those when A is perfectly reliable (i.e., set 6 in Table 4.5). Comparing the remaining sets in Table 4.6, the sets with greater values of r_A have greater probabilities of states G and D.

4.5.2.3 Controller with Imperfect Coverage

Table 4.7 lists system state probabilities of the example cloud-RAID 5 for different combinations of c_A, r_A, and s_A with $\lambda_A = 0.0001/h$, $t_A = 1000$ h and disk state probabilities of set 6 in Table 4.2.

Set 1 in Table 4.7 shows the highest system UF state probability among all the sets, because the controller fault is always uncovered ($s_A = 1$). For sets 1 through 4, the controller fault is either transient or uncovered; a set with higher s_A leads to a higher system UF state probability. For sets 1, 5, 6, and 7 having $r_A = 0$ (meaning that the controller has no transient faults), because the controller itself is a single-point failure (its CF will crash the system just like its UF), the system under those four sets of parameters has the same G or D state probability, and a set with a higher s_A (c_A) leads to a higher system UF (CF) state probability. The same conclusion can be drawn from data under sets (2, 10), (3, 9, 13, 15), (4, 8, 11, 16), or (12, 14) with the same value of r_A (thus, the same sum of c_A and s_A). For sets 8 through 10 having $s_A = 0$ (meaning that the controller has no uncovered faults), a set with a higher value of r_A leads to a higher system G or D state probability.

TABLE 4.7

System State Probabilities (A Has UF)

No	c_A	r_A	s_A	G	D	CF	UF
1	0	0	1	0.792229	0.002182	0.064339	0.141249
2	0	0.7	0.3	0.850553	0.002343	0.069075	0.078029
3	0	0.5	0.5	0.833889	0.002297	0.067722	0.096092
4	0	0.3	0.7	0.817225	0.002251	0.066369	0.114155
5	0.7	0	0.3	0.792229	0.002182	0.130953	0.074636
6	0.5	0	0.5	0.792229	0.002182	0.111920	0.093669
7	0.3	0	0.7	0.792229	0.002182	0.092888	0.112701
8	0.7	0.3	0	0.817225	0.002251	0.132983	0.047541
9	0.5	0.5	0	0.833889	0.002297	0.115303	0.048511
10	0.3	0.7	0	0.850553	0.002343	0.097624	0.049480
11	0.5	0.3	0.2	0.817225	0.002251	0.113950	0.066574
12	0.5	0.2	0.3	0.808893	0.002228	0.113273	0.075606
13	0.3	0.5	0.2	0.833889	0.002297	0.096271	0.067543
14	0.3	0.2	0.5	0.808893	0.002228	0.094241	0.094638
15	0.2	0.5	0.3	0.833889	0.002297	0.086755	0.077059
16	0.2	0.3	0.5	0.817225	0.002251	0.085401	0.095123

TABLE 4.8
System State Probabilities

t_A	G	D	CF	UF
0	0.875549	0.002411	0.071105	0.050934
1,000	0.833889	0.002297	0.096271	0.067543
2,000	0.796194	0.002193	0.119042	0.082572
3,000	0.762086	0.002099	0.139645	0.096170
5,000	0.703298	0.001937	0.175157	0.119608
10,000	0.598823	0.001649	0.238268	0.161260
15,000	0.535455	0.001475	0.276547	0.186524
20,000	0.497021	0.001369	0.299764	0.201847

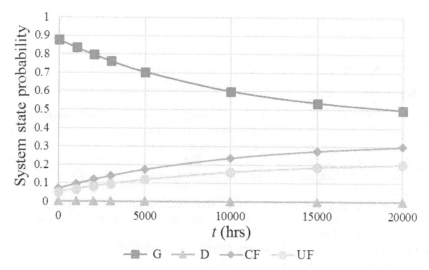

FIGURE 4.13
System state probabilities.

Table 4.8 lists the system state probabilities for several different mission times t_A (in hours) under $\lambda_A = 0.0001/h$, $c_A = 0.3$, $r_A = 0.5$, $s_A = 0.2$. These results are graphically represented in Figure 4.13. With the mission time proceeding, the system G and D state probabilities continuously decrease while the system CF and UF state probabilities increase.

4.6 Conclusion and Future Work

Reliability modeling of multi-state cloud-RAID 5 systems subject to ELC is presented, encompassing Markov models for the disk-level and MMDD-based combinatorial procedure for system-level state probability analysis.

Effects of functional dependence behavior between the RAID controller and disks are addressed. Numerical results demonstrate effects of each coverage factor on the disk or system state probabilities. Moreover, these results show that failure to consider imperfect fault coverage leads to inaccurate disk or system state probabilities, thus misleading system design, maintenance, and optimization activities.

While the cloud-RAID 5 system is used as an illustrating case study, the proposed methodology is applicable to other levels of cloud-RAID storage systems such as cloud-RAID 6. In the future, we will investigate reliability analysis of cloud-RAID systems subject to other types of coverage models; for instance, fault-level coverage where the coverage probability relies on the number of component faults occurring in a particular group within a certain recovery window (Amari et al., 2008).

References

Akers, J., R. Bergman, S. V. Amari, and L. Xing. 2008. Analysis of multi-state systems using multi-valued decision diagrams. In: *Proceedings of Annual Reliability and Maintainability Symposium (RAMS)*, 347–353 Las Vegas, NV.

Amari, S. V., J. B. Dugan, and R. B. Misra. 1999. A separable method for incorporating imperfect fault-coverage into combinatorial models. *IEEE Trans. Reliab.* 48:267–274.

Amari, S. V., A. Myers, A. Rauzy, and K. Trivedi. 2008. Imperfect coverage models: Status and trends. In: *Handbook of Performability Engineering*, ed. K. B. Misra, 321–348. London: Springer.

Bausch, F. 2014. Cloud-RAID concept. http://blog.fbausch.de/cloudraid-3-concept/ (accessed December 2017).

Chang, Y. R., S. V. Amari, and S. Y. Kuo. 2005. OBDD-based evaluation of reliability and importance measures for multi-state systems subject to imperfect fault coverage. *IEEE Trans. Dependable Secure Comput.* 2:336–347.

Deng, J., S. C. H. Huang, Y. S. Han, and J. H. Deng. 2010. Fault tolerant and reliable computation in cloud computing. In *Proceedings of IEEE Globecom workshops*, 1601–1605. Miami, FL.

Erl, T., R. Puttini, and Z. Mahmood. 2013. *Cloud Computing Concepts, Technology and Architecture*. The Prentice Hall Service Technology Series. Prentice Hall. New Jersey, NJ.

Fitch, D. and H. Xu. 2013. A RAID-based secure and fault-tolerant model for cloud information storage. *Int. J. Software Eng. Knowl. Eng.* 23:627–654.

Gilroy, M. and J. Irvine. 2006. RAID 6 hardware acceleration. In: *Proceedings of International Conference on Field Programmable Logic and Applications* 1–6. Madrid, Spain.

Jin, T., Y. Yu, and L. Xing. 2009. Reliability analysis of RAID systems using repairable k-out-of-n modeling techniques. In: *Proceedings of International Conference on the Interface between Statistics and Engineering*, 1–8. Beijing, China.

Jin, T., L. Xing, and Y. Yu. 2011. A hierarchical Markov reliability model for data storage systems with media self-recovery. *Int. J. Reliab. Qual. Saf. Eng.* 18:25–41.

Junges. S., D. Guck, J. P. Katoen, and M. Stoelinga. 2016. Uncovering dynamic fault trees. In: *Proceedings of 46th Annual IEEE/IFIP International Conference on Dependable Systems and Networks*, 299–310. Toulouse, France.

Levitin, G. and S. V. Amari. 2008. Multi-state systems with multi-fault coverage. *Reliab. Eng. Syst. Saf.* 93:1730–1739.

Levitin, G. and S. V. Amari. 2009. Three types of fault coverage in multi state systems. In: *Proceedings of International Conference on Reliability Maintainability and Safety*, 122–127. Chengdu, China.

Li, Q. and C. Mao. 2016. Considering testing-coverage and fault removal efficiency subject to the random field environments with imperfect debugging in software reliability assessment. In: *Proceedings of IEEE 27th International Symposium on Software Reliability Engineering Workshops*, 257–263. Ottawa, ON.

Li, W., Y. Yang, and J. Chen. 2012. A cost-effective mechanism for cloud data reliability management based on proactive replica checking. In: *Proceedings of IEEE/ACM International Symposium on Cluster, Cloud and Grid Computing*, 564–571. Ottawa, ON.

Li, W., Y. Yang, and D. Yuan. 2016. Ensuring cloud data reliability with minimum replication by proactive replica checking. *IEEE Trans. Comput.* 65:1494–1506.

Lin, C. Y. and W. G. Tzeng. 2014. Game-theoretic strategy analysis for data reliability management in cloud storage systems. In: *International Conference on Software Security and Reliability*, 187–195. San Francisco, CA.

Liu, Q. and L. Xing. 2015a. Reliability modeling of cloud-RAID-6 storage system. *Int. J. Future Comput. Commun.* 4:415–420.

Liu, Q. and L. Xing. 2015b. Hierarchical reliability analysis of multi-state cloud-RAID storage system. In: *Proceedings of International Conference on Quality, Reliability, Risk, Maintenance, and Safety Engineering*, 1–7. Beijing, China.

Lu, Y. and S. Xia. 2015. Novel erasure codes with repair optimality for cloud storage. In: *International Conference on Information and Communications Technologies*, 1–5. Xi'an, China.

Luo, S., M. Hou, S. Zhan, M. Lyu, and M. Li. 2017. Consistency maintenance in replication: A novel strategy based on diamond topology in cloud storage. *Chin. J. Electron.* 26:192–198.

Mandava, L. and L. Xing. 2017. Reliability analysis of cloud-RAID 6 with imperfect fault coverage. *Int. J. Performab. Eng.* 13:289–297.

Mandava, L., L. Xing, and Z. Pan. 2016. Imperfect coverage analysis for cloud-RAID 5. In: *Engineering Asset Management*, eds. M. Zuo, L. Ma, J. Mathew, and H. Z. Huang, 207–220. Lecture Notes in Mechanical Engineering. Cham: Springer.

Myers, A. 2010. *Complex System Reliability*. Springer Series in Reliability Engineering. 2nd edn. Springer-Verlag, London, UK.

Myers, A. and A. Rauzy. 2008. Efficient reliability assessment of redundant systems subject to imperfect fault coverage using binary decision diagrams. *IEEE Trans. Reliab.* 57:336–348.

Patterson, D., P. Chen, G. Gibson, and R. H. Katz. 1989. Introduction to redundant arrays of inexpensive disks (RAID). In: *Proceedings of IEEE Computer Society International Conference: Intellectual Leverage, Digest of Papers*, 112–117. San Francisco, CA.

Pham, H., S. V. Amari, and R. B. Misra. 1996. Reliability analysis of k-out-of-n systems with partially reparable multi-state components. *Microelectron. Reliab.* 36:1407–1415.

Shrestha, A., L. Xing, and Y. Dai. 2007. MBDD versus MMDD for multistate systems analysis. In: *Proceedings of IEEE International Symposium on Dependable, Autonomic and Secure Computing*, 172–180. Colombia, MD.

Shrestha, A., L. Xing, and S. V. Amari. 2010. Reliability and sensitivity analysis of imperfect coverage multi-state systems. In: *Proceedings of Reliability and Maintainability Symposium (RAMS)*, 1–6. San Jose, CA.

Xiang, J., F. Machida, K. Tadano, and Y. Maeno. 2015. An imperfect fault coverage model with coverage of irrelevant components. *IEEE Trans. Reliab.* 64:320–332.

Xing, L. and S. V. Amari. 2015. *Binary Decision Diagrams and Extensions for System Reliability Analysis*. Wiley-Scrivener. Hoboken, NJ.

Xing, L. and Y. Dai. 2009. A new decision diagram based method for efficient analysis on multi-state systems. *IEEE Trans. Depend. Secure Comput.* 6:161–174.

Xing, L., B. A. Morrissette, and J. B. Dugan. 2014. Combinatorial reliability analysis of imperfect coverage systems subject to functional dependence. *IEEE Trans. Reliab.* 63:367–382.

Yin, J., Y. Tang, S. Deng, Y. Li, W. Lo, K. Dong, A. Y. Zomaya, and C. Pu. 2017. ASSER: An efficient, reliable, and cost-effective storage scheme for object-based cloud storage systems. *IEEE Trans. Comput.* 66:1326–1340.

Zhang, R., C. Lin, K. Meng, and L. Zhu. 2013a. BEC: A reliable and efficient mechanism for cloud storage service. In: *Proceedings of IEEE International Conference on Communication Technology*, 717–721. Guilin, China.

Zhang, R., C. Lin, K. Meng, and L. Zhu. 2013b. A modeling reliability analysis technique for cloud storage system. In: *Proceedings of the 15th IEEE International Conference on Communication Technology*, 32–36. Guilin, China.

5

L_z-Transform Approach for Comparison of Different Schemas of Ships' Diesel-Electric Multi-Power Source Traction Drives

Ilia Frenkel, Igor Bolvashenkov, Lev Khvatskin, and Anatoly Lisnianski

CONTENTS

5.1 Introduction 83
5.2 L_z-Transform Method 84
5.3 Multi-State Models of the Multi-Power Source Traction Drives 86
    5.3.1 Systems Description 86
    5.3.2 Elements Description 89
    5.3.3 Diesel-Generators and Electric Motors Subsystems 90
        5.3.3.1 Diesel-Generator Subsystem 90
        5.3.3.2 Electric Motors Subsystem 92
    5.3.4 Multi-State Model for Ship's Conventional Diesel-Electric Traction Drive 93
    5.3.5 Multi-State Model for Ship's Alternative Diesel-Electric Traction Drive 95
    5.3.6 Calculation Reliability Indices of Multi-Power Source Traction Drives 99
5.4 Concluding Remarks 103
References 103

5.1 Introduction

Vehicle traction drives are safety-critical systems. In this regard, they are subject to stringent requirements for safety, survivability, and sustainability. Especially important is the implementation of these requirements for Arctic icebreaking ships. In this chapter, an attempt is made to estimate in terms of the above parameters two topologies of hybrid-electric traction drives of an Amguema-type Arctic icebreaking ship, described in detail in [1].

Due to the system's nature, a fault in a single unit has only a partial effect on the entire power performance. A partial failure of the multi-power source traction drive leads to partial system failure (reduction of output nominal power), as well as multiple consecutive failures, to complete system failure. So, the ship's diesel-electric multi-power source traction drive can be regarded as a multi-state system (MSS), where components, as well as the whole system, are considered to have a finite number of states associated with various performance rates [1–3]. The system's performance rate (output nominal power) is viewed as a discrete-state continuous-time stochastic process. Such models, even in simple settings, are quite complex since they may have a few hundreds of states. Therefore, not only the construction of such a model, but also the solution of the associated system of differential equations via a straightforward Markov method is very complicated.

In recent years, a special technique known as the L_z-transform was proposed and investigated [5] for discrete-state continuous-time Markov processes. This approach is an extension of the universal generating function (UGF) technique that was proposed by Ushakov [7] that has been extensively implemented for the analysis of reliability of MSSs. The L_z-transform has turned out to be a powerful and highly efficient tool in availability analysis for MSS for both constant and variable demand [4, 5, 8]. Note that the above technique has great applicability for numerous structure functions [5].

In this chapter, the L_z-transform is applied to analyze a real MSS multi-power source traction drive that functions under different weather conditions, and its availability, power performance, and power performance deficiency are investigated. It is established that the implementation of the L_z-transform simplifies the system's availability computation considerably as compared with the standard Markov model.

5.2 L_z-Transform Method

To analyze MSS behavior, the characteristics of system elements must be known. Any system element j can have k_j different states, which correspond to the elements' performances g_{ji}, defined by the set $\mathbf{g}_j = \{g_{j1},...,g_{jk_j}\}$, $j = \{1,...,n\}$; $i = \{1,2,...,k_j\}$. The performance stochastic processes $G_j(t) \in \mathbf{g}_j$ and the system structure function $G(t) = f(G_1(t),...,G_n(t))$ fully define the MSS model. To determine the L_z-transform of the entire MSS's output performance Markov process, the following steps must be performed.

Firstly, for each system element j, there must be built a model of the Markov performance stochastic process, represented by the expression

$$G_j(t) = \{\mathbf{g}_j, \mathbf{A}_j, \mathbf{p}_{j0}\}$$

where:

$$\mathbf{g}_j$$

is the set of possible component states

$$\mathbf{A}_j = \left(a_{lm}^{(j)}(t)\right), \quad l, m = 1, \ldots, k; j = 1, \ldots n$$

is the transition intensities matrix

$$\mathbf{p}_{j0} = \left[p_{10}^{(j)} = \Pr\{G_j(0) = g_{10}\}, \ldots, p_{k_j 0}^{(j)} = \Pr\{G_j(0) = g_{k_j 0}\}\right]$$

is the probability distribution of the initial state and the system of Kolmogorov differential equations [6] for determination of the probabilities of each state is $p_{ji}(t) = \Pr\{G_j(t) = g_{ji}\}$, $i = 1, \ldots k_j, j = 1, \ldots, n$ with initial conditions $\mathbf{p}_{j0}$.

The L_z-transform of a discrete-state continuous-time (DSCT) Markov process $G_j(t)$ for each component j can be written in the following form:

$$L_Z\{G_j(t)\} = \sum_{i=1}^{k_j} p_{ji}(t) z^{g_{ji}}. \tag{5.1}$$

For the determination of the L_z-transform of the entire MSS's output performance Markov process $G(t)$, we apply Ushakov's universal generating operator [7] to each of the individual L_Z- transforms $L_Z\{G_j(t)\}$ over all time points $t \geq 0$.

$$L_Z\{G(t)\} = \Omega_f\{L_Z[G_1(t)], \ldots, L_Z[G_n(t)]\} = \sum_{i=1}^{K} p_i(t) z^{g_i} \tag{5.2}$$

Then, considering all terms with nonnegative power, we obtain the corresponding MSS mean instantaneous availability:

$$A(t) = \sum_{g_i \geq w} p_i(t). \tag{5.3}$$

The use of terms with positive powers, together with the corresponding performance rates, provides the expression for the MSS's mean instantaneous performance:

$$E(t) = \sum_{g_i > 0} p_i(t) g_i. \tag{5.4}$$

Finally, the expression for the instantaneous performance deficiency $D(t)$ for any t and for the case of constant demand w is given by:

$$D(t) = \sum_{i=1}^{K} p_i(t) \cdot \max(w - g_i, 0). \tag{5.5}$$

5.3 Multi-State Models of the Multi-Power Source Traction Drives

5.3.1 Systems Description

We analyze a conventional diesel-electric power drive, used in Amguema-type Arctic cargo ships, based on a direct electric propulsion system, comparing it with an alternative diesel-electric power drive. Structures of ships' diesel-electric traction drives are shown in Figure 5.1.

The power performance of both complete systems is 5500 kW. Depending on the ice conditions, the amount of cargo, and other conditions of navigation, the ship's propulsion system operates with a different number of diesel-generators and electric propulsion motors. It realizes the required value of the performance and, as a consequence, the high level of survivability of the ship with the possible occurrence of critical failures of power equipment.

The structure of the conventional schema of a ship's diesel-electric traction drive is shown in Figure 5.1a. The system consists of a diesel-generators subsystem, a switchboard, an electrical energy converter, an electric motors subsystem, and a fixed pitch propeller.

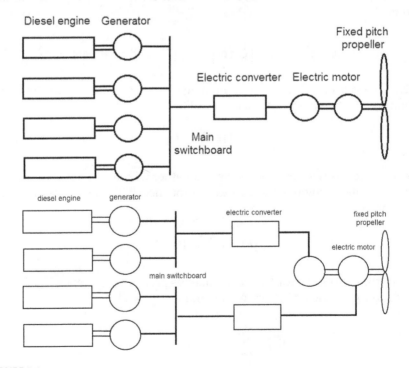

FIGURE 5.1
(a) Structure of a ship's conventional diesel-electric traction drive. (b) Structure of an ship's alternative diesel-electric traction drive.

The diesel-generator subsystem consists of four diesel-generators. Each diesel-generator consists of a diesel engine and a generator. The power-generating performance of each diesel-generator is 1375 kW. Therefore, connecting diesel-generators in parallel supports the nominal performance required for the functioning of the whole system.

The main switchboard device and the electrical energy converter have nominal performance.

The electric motors subsystem consists of two motors with a performance of 2750 kW. Therefore, connecting motors in parallel supports the nominal performance required for the functioning of the whole system.

In ships' diesel-electric power drives with a fixed pitch propeller, the dimensioning of the electric machines has to be calculated accurately to estimate sufficient available propulsion power, which is directly determined by the required values of operational power and necessary additional power in the case of heavy weather or icy conditions in the area of navigation.

The possible structures of the Arctic ship's propulsion system with a different number of diesel-generators and main traction motors are determined by the operating conditions of the Arctic ship and the ice and temperature conditions.

Typical operational modes of Arctic cargo ships are as follows:

- Navigation with an icebreaker in heavy ice and navigation without an icebreaker in solid ice requires 75% of generated power.
- Navigation in open water depends on the required velocity and requires 50% of generated power.

The reliability block diagram of a ship's conventional diesel-electric traction drive is presented in Figure 5.2.

The structure of an alternative schema of a ship's diesel-electric traction drive is shown in Figure 5.1b. The system consists of two similar subsystems connected in parallel; each contains a diesel-generators subsystem, main switchboards, electrical energy converters and electric motors, and a fixed pitch propeller.

The performance of the whole system is 5500 kW. The performance of each subsystem is 2750 kW. Therefore, connecting two subsystems in parallel supports the nominal performance required for the functioning of the whole system.

Each diesel-generator consists of a diesel engine and a generator. The performance of each diesel-generator is 1375 kW.

Each main switchboard device, each electrical energy converter, and each electric motor has a nominal 2750 kW performance.

The reliability block diagram of the alternative schema of a ship's diesel-electric traction drive is presented in Figure 5.3.

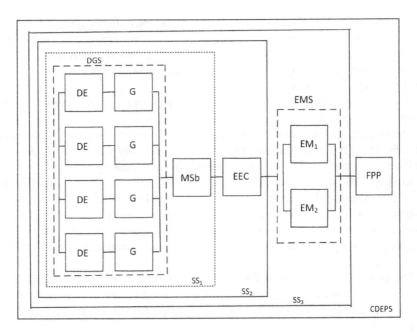

FIGURE 5.2
Reliability block diagram of a ship's conventional diesel-electric traction drive.

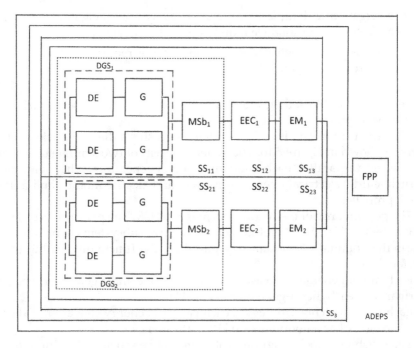

FIGURE 5.3
Reliability block diagram of an alternative schema of a ship's diesel-electric traction drive.

5.3.2 Elements Description

All system elements have two states (fully working and fully failed). To calculate the probabilities of each state, we build the state space diagram presented in Figure 5.4 and the system of differential equations given in Equation 5.6:

$$\begin{cases} \dfrac{dp_{i1}(t)}{dt} = -\lambda_i p_{i1}(t) + \mu_i p_{i2}(t), \\ \dfrac{dp_{i2}(t)}{dt} = \lambda_i p_{i1}(t) - \mu_i p_{i2}(t). \end{cases} \quad (5.6)$$

where i = DE, G, MSb, MSb$_1$, MSb$_2$, EEC, EEC$_1$, EEC$_2$, EM, EM$_1$, EM$_2$, FPP.

Initial conditions are: $p_{i1}(0) = 1$; $p_{i2}(0) = 0$.

Note that to solve the above system of differential equations and obtain the required probabilities $p_{i1}(t), p_{i2}(t)$ (i = DE, G, MSb, EEC, EM, FPP), we used MATLAB® software.

Therefore, for such system elements, the output performance stochastic processes can be obtained as follows:

- Ship's conventional diesel-electric traction drive

For i = DE, G	For i = EM$_1$, EM$_2$	For i = MSb, EEC, EM, FPP
$\begin{cases} \mathbf{g}_i = \{g_{i1}, g_{i1}\} = \{1375, 0\} \\ \mathbf{p}_i(t) = \{p_{i1}(t), p_{i1}(t)\}. \end{cases}$	$\begin{cases} \mathbf{g}_i = \{g_{i1}, g_{i1}\} = \{2750, 0\} \\ \mathbf{p}_i(t) = \{p_{i1}(t), p_{i1}(t)\}. \end{cases}$	$\begin{cases} \mathbf{g}_i = \{g_{i1}, g_{i1}\} = \{5500, 0\} \\ \mathbf{p}_i(t) = \{p_{i1}(t), p_{i1}(t)\}. \end{cases}$

- Ship's alternative diesel-electric traction drive

For i = DE, G	For i = EEC$_1$, EEC$_2$, EEC$_1$, EEC$_2$, EM$_1$, EM$_2$	For i = FPP
$\begin{cases} \mathbf{g}_i = \{g_{i1}, g_{i1}\} = \{1375, 0\} \\ \mathbf{p}_i(t) = \{p_{i1}(t), p_{i1}(t)\}. \end{cases}$	$\begin{cases} \mathbf{g}_i = \{g_{i1}, g_{i1}\} = \{2750, 0\} \\ \mathbf{p}_i(t) = \{p_{i1}(t), p_{i1}(t)\}. \end{cases}$	$\begin{cases} \mathbf{g}_i = \{g_{i1}, g_{i1}\} = \{5500, 0\} \\ \mathbf{p}_i(t) = \{p_{i1}(t), p_{i1}(t)\}. \end{cases}$

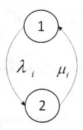

FIGURE 5.4
State space diagram.

Sets $\mathbf{g}_i, \mathbf{p}_i(t)$ (i = DE, G, MSb, MSb$_1$, MSb$_2$, EEC, EEC$_1$, EEC$_2$, EM, EM$_1$, EM$_2$, FPP) define L_z-transforms for each element as follows:

Diesel engine:

$$L_z\left\{g^{DE}(t)\right\} = p_1^{DE}(t) z^{g_1^{DE}} + p_2^{DE}(t) z^{g_2^{DE}} = p_1^{DE}(t) z^{1375} + p_2^{DE}(t) z^0. \quad (5.7)$$

Generator:

$$L_z\left\{g^{G}(t)\right\} = p_1^{G}(t) z^{g_1^{G}} + p_2^{G}(t) z^{g_2^{G}} = p_1^{G}(t) z^{1375} + p_2^{G}(t) z^0. \quad (5.8)$$

Main switchboards:

$$L_z\left\{g^{MSb}(t)\right\} = p_1^{MSb}(t) z^{g_1^{MSb}} + p_2^{MSb}(t) z^{g_2^{MSb}} = p_1^{MSb}(t) z^{5500} + p_2^{MSb}(t) z^0$$

$$L_z\left\{g^{MSb_1}(t)\right\} = p_1^{MSb_1}(t) z^{g_1^{MSb_1}} + p_2^{MSb_1}(t) z^{g_2^{MSb_1}} = p_1^{MSb_1}(t) z^{2750} + p_2^{MSb_1}(t) z^0 \quad (5.9)$$

$$L_z\left\{g^{MSb_2}(t)\right\} = p_1^{MSb_2}(t) z^{g_1^{MSb_2}} + p_2^{MSb_2}(t) z^{g_2^{MSb_2}} = p_1^{MSb_2}(t) z^{2750} + p_2^{MSb_2}(t) z^0$$

Electrical energy converters:

$$L_z\left\{g^{EEC}(t)\right\} = p_1^{EEC}(t) z^{g_1^{EEC}} + p_2^{EEC}(t) z^{g_2^{EEC}} = p_1^{EEC}(t) z^{5500} + p_2^{EEC}(t) z^0$$

$$L_z\left\{g^{EEC_1}(t)\right\} = p_1^{EEC_1}(t) z^{g_1^{EEC_1}} + p_2^{EEC_1}(t) z^{g_2^{EEC_1}} = p_1^{EEC_1}(t) z^{2750} + p_2^{EEC_1}(t) z^0 \quad (5.10)$$

$$L_z\left\{g^{EEC_2}(t)\right\} = p_1^{EEC_2}(t) z^{g_1^{EEC_2}} + p_2^{EEC_2}(t) z^{g_2^{EEC_2}} = p_1^{EEC_2}(t) z^{2750} + p_2^{EEC_2}(t) z^0$$

Electric motors:

$$L_z\left\{g^{EM_1}(t)\right\} = p_1^{EM_1}(t) z^{g_1^{EM_1}} + p_2^{EM_1}(t) z^{g_2^{EM_1}} = p_1^{EM_1}(t) z^{2750} + p_2^{EM_1}(t) z^0$$

$$L_z\left\{g^{EM_2}(t)\right\} = p_1^{EM_2}(t) z^{g_1^{EM_2}} + p_2^{EM_2}(t) z^{g_2^{EM_2}} = p_1^{EM_2}(t) z^{2750} + p_2^{EM_2}(t) z^0 \quad (5.11)$$

Fixed pitch propeller:

$$L_z\left\{G^{FPP}(t)\right\} = p_1^{FPP}(t) z^{g_1^{FPP}} + p_2^{FPP}(t) z^{g_2^{FPP}} = p_1^{FPP}(t) z^{5500} + p_2^{FPP}(t) z^0. \quad (5.12)$$

5.3.3 Diesel-Generators and Electric Motors Subsystems

5.3.3.1 Diesel-Generator Subsystem

A diesel-generator subsystem for a ship's conventional diesel-electric traction drive consists of four pairs of identical diesel engines and generators connected in parallel. The diesel-generator subsystem for an ship's alternative diesel-electric traction drive consists of two pairs of identical diesel engines and generators connected in parallel. Each diesel engine and each generator are two-state devices: the power performance of a fully operational state is 1375 kW and a total failure corresponds to a capacity of 0.

Using the composition operator Ω_{fser}, the L_z-transform $L_z\{G^{DG}(t)\}$ is obtained for each pair of identical diesel engines and generators, connected in series, where the powers of z are found as the minima of powers of corresponding terms:

$$L_z\{G^{DG}(t)\} = \Omega_{fser}\left(g^{DE}(t), g^G(t)\right)$$
$$= p_1^{DE}(t)p_1^G(t)z^{1375} + \left(p_1^{DE}(t)p_2^G(t) + p_{i2}^{DE}(t)\right)z^0. \quad (5.13)$$

If

$$p_1^{DG}(t) = p_1^{DE}(t)p_1^G(t)$$
$$p_2^{DG}(t) = p_1^{DE}(t)p_2^G(t) + p_2^{DE}(t)$$

then the L_z-transform for the diesel-generator subsystem is given by

$$L_z\{G^{DG}(t)\} = p_1^{DG}(t)z^{1375} + p_2^{DG}(t)z^0. \quad (5.14)$$

Using again the Ω_{fpar} operator for four diesel-generators connected in parallel, the L_z-transform $L_z\{G^{SysDG}(t)\}$ is obtained for whole conventional diesel-generator subsystem as follows:

$$L_z\{G^{DGS}(t)\} = \Omega_{fpar}\left(L_z\{G^{DG}(t)\}, L_z\{G^{DG}(t)\}, L_z\{G^{DG}(t)\}, L_z\{G^{DG}(t)\}\right). \quad (5.15)$$

Using notations

$$P_1^{DGS}(t) = \{p_1^{DG}(t)\}^4;$$
$$P_2^{DGS}(t) = 4 \cdot \{p_1^{DG}(t)\}^3 p_2^{DG}(t);$$
$$P_3^{DGS}(t) = 6 \cdot \{p_1^{DG}(t)\}^2 \{p_2^{DG}(t)\}^2;$$
$$P_4^{DGS}(t) = 4 \cdot p_1^{DG}(t)\{p_2^{DG}(t)\}^3;$$
$$P_5^{DGS}(t) = \{p_2^{DG}(t)\}^4.$$

the L_z-transform for the entire diesel-generator subsystem is given by

$$L_z\{G^{DGS}(t)\} = P_1^{DGS}(t)z^{5500} + P_2^{DGS}(t)z^{4125} + P_3^{DGS}(t)z^{2750}$$
$$+ P_4^{DGS}(t)z^{1375} + P_4^{DGS}(t)z^0. \quad (5.16)$$

Now, applying Ω_{fpar} again for two similar diesel-generators connected in parallel, the L_z-transform $L_z\{G^{DGS_i}(t)\}, i=1,2$ for the alternative diesel-generator subsystem is obtained:

$$L_z\{G^{DGS_i}(t)\} = \Omega_{fpar}\left(L_z\{G^{DG}(t)\}, L_z\{G^{DG}(t)\}\right), \quad i=1,2. \quad (5.17)$$

Using notations

$$P_1^{DGS_i}(t) = \{p_1^{DG}(t)\}^2;$$
$$P_2^{DGS_i}(t) = 2p_1^{DG}(t)p_2^{DG}(t);$$
$$P_3^{DGS_i}(t) = \{p_2^{DG}(t)\}^2.$$

the L_z-transform for the diesel-generator subsystems is given as follows:

$$L_z\{G^{DGS_i}(t)\} = P_1^{DGS_i}(t)z^{2750} + P_2^{DGS_i}(t)z^{1375} + P_3^{DGS_i}(t)z^0, \quad i=1,2. \quad (5.18)$$

5.3.3.2 Electric Motors Subsystem

The electric motors subsystem for a conventional diesel-electric traction drive consists of two electric motors connected in parallel. Each electric motor is a two-state device: the power performance of a fully operational state is 2750 kW and a total failure corresponds to a capacity of 0.

Applying the Ω_{fpar} operator for two electric motors, connected in parallel, we obtain the L_z-transform $L_z\{G^{EMS}(t)\}$ for the whole electric motors subsystem as follows:

$$L_z\{G^{EMS}(t)\} = \Omega_{fpar}\left(L_z\{G^{EM_1}(t)\}, L_z\{G^{EM_2}(t)\}\right). \quad (5.19)$$

Using notations

$$P_1^{EMS}(t) = p_1^{EM_1}(t)p_2^{EM_2}(t);$$
$$P_2^{EMS}(t) = p_1^{EM_1}(t)p_2^{EM_2}(t) + p_2^{EM_1}(t)p_1^{EM_2}(t);$$
$$P_3^{EMS}(t) = p_2^{EM_1}(t)p_2^{EM_2}(t);$$

the L_z-transform of the entire diesel-generator subsystem is given below:

$$L_z\{G^{EMS}(t)\} = P_1^{EMS}(t)z^{5500} + P_2^{EMS}(t)z^{2750} + P_3^{EMS}(t)z^0. \quad (5.20)$$

5.3.4 Multi-State Model for Ship's Conventional Diesel-Electric Traction Drive

As was evident from Figure 5.2, the multi-state model for the multi-power source traction drive may be presented as a diesel-generator subsystem connected in series, a main switchboard, an electrical energy converter and electric motor subsystem, and a fixed pitch propeller. Therefore, the L_z-transform for the whole system is as follows:

$$L_z\{G^{CDEPS}(t)\} = \Omega_{f_{ser}}\left(L_z\{G^{DGS}(t)\}, L_z\{G^{MSb}(t)\}, L_z\{G^{EEC}(t)\}, \right.$$
$$\left. L_z\{G^{EMS}(t)\}, L_z\{G^{FPP}(t)\}\right) \quad (5.21)$$

Using the recursive derivation approach [5], we will calculate the L_z-transform of the entire system:

$$L_z\{G^{SS_1}(t)\} = \Omega_{f_{ser}}\left(L_z\{G^{DGS}(t)\}, L_z\{G^{MSb}(t)\}\right)$$

$$L_z\{G^{SS_2}(t)\} = \Omega_{f_{ser}}\left(L_z\{G^{SS_1}(t)\}, L_z\{G^{EEC}(t)\}\right)$$

$$L_z\{G^{SS_3}(t)\} = \Omega_{f_{ser}}\left(L_z\{G^{SS_2}(t)\}, L_z\{G^{EMS}(t)\}\right)$$

$$L_z\{G^{CDEPS}(t)\} = \Omega_{f_{ser}}\left(L_z\{G^{SS_3}(t)\}, L_z\{G^{FPP}(t)\}\right)$$

Using the composition operator $\Omega_{f_{ser}}$ for subsystems and elements connected in series, we obtain the following L_z-transforms:

- L_z-transforms for SS_1 subsystem

$$L_z\{G^{SS_1}(t)\} = \Omega_{f_{ser}}\left(L_z\{G^{DGS}(t)\}, L_z\{G^{MSb}(t)\}\right)$$
$$= \Omega_{f_{ser}}\left(P_1^{DG}(t)z^{5500} + P_2^{DG}(t)z^{4125} + P_3^{DG}(t)z^{2750}\right.$$
$$\left. + P_4^{DG}(t)z^{1375} + P_4^{DG}(t)z^0, p_1^{MSb}(t)z^{5500} + p_2^{MS}(t)z^0\right)$$
$$= P_1^{SS_1}(t)z^{5500} + P_2^{SS_1}(t)z^{4125} + P_3^{SS_1}(t)z^{2750} + P_4^{SS_1}(t)z^{1375} + P_5^{SS_1}(t)z^0. \quad (5.22)$$

where:

$$P_1^{SS_1}(t) = P_1^{DGS}(t) p_1^{MSb}(t),$$

$$P_2^{SS_1}(t) = P_2^{DGS}(t) p_1^{MSb}(t),$$

$$P_3^{SS_1}(t) = P_3^{DGS}(t) p_1^{MSb}(t),$$

$$P_4^{SS_1}(t) = P_4^{DGS}(t) p_1^{MSb}(t),$$

$$P_5^{SS_1}(t) = P_5^{DGS}(t) p_1^{MSb}(t) + p_2^{MSb}(t).$$

- L_z-transforms for SS$_2$ subsystem

$$\begin{aligned}L_z\{G^{SS_2}(t)\} &= \Omega_{f_{ser}}\left(L_z\{G^{SS_1}(t)\}, L_z\{G^{EEC}(t)\}\right) \\ &= \Omega_{f_{ser}}\Big(P_1^{SS_1}(t) z^{5500} + P_2^{SS_1}(t) z^{4125} + P_3^{SS_1}(t) z^{2750} \\ &\quad + P_4^{SS_1}(t) z^{1375} + P_4^{SS_1}(t) z^0, p_1^{EEC}(t) z^{5500} + p_2^{EEC}(t) z^0\Big) \\ &= P_1^{SS_2}(t) z^{5500} + P_2^{SS_2}(t) z^{4125} + P_3^{SS_2}(t) z^{2750} + P_4^{SS_2}(t) z^{1375} + P_5^{SS_2}(t) z^0. \end{aligned} \quad (5.23)$$

where:

$$P_1^{SS_2}(t) = P_1^{SS_1}(t) p_1^{EEC}(t),$$

$$P_2^{SS_2}(t) = P_2^{SS_1}(t) p_1^{EEC}(t),$$

$$P_3^{SS_2}(t) = P_3^{SS_1}(t) p_1^{EEC}(t),$$

$$P_4^{SS_2}(t) = P_4^{SS_1}(t) p_1^{EEC}(t),$$

$$P_5^{SS_2}(t) = P_5^{SS_1}(t) p_1^{EEC}(t) + p_2^{EEC}(t).$$

- L_z-transforms for SS$_3$ subsystem

$$\begin{aligned}L_z\{G^{SS_3}(t)\} &= \Omega_{f_{ser}}\left(L_z\{G^{SS_2}(t)\}, L_z\{G^{EMS}(t)\}\right) \\ &= \Omega_{f_{ser}}\Big(P_1^{SS_2}(t) z^{5500} + P_2^{SS_2}(t) z^{4125} + P_3^{SS_2}(t) z^{2750} + P_4^{SS_2}(t) z^{1375} \\ &\quad + P_4^{SS_2}(t) z^0, p_1^{EMS}(t) z^{5500} + p_2^{EMS}(t) z^{2750} + p_3^{EMS}(t) z^0\Big) \\ &= P_1^{SS_3}(t) z^{5500} + P_2^{SS_3}(t) z^{4125} + P_3^{SS_3}(t) z^{2750} + P_4^{SS_3}(t) z^{1375} + P_5^{SS_3}(t) z^0. \end{aligned} \quad (5.24)$$

where:

$$P_1^{SS_3}(t) = P_1^{SS_2}(t)P_1^{EMS}(t),$$

$$P_2^{SS_3}(t) = P_2^{SS_1}(t)P_1^{EMS}(t),$$

$$P_3^{SS_3}(t) = P_1^{SS_1}(t)P_2^{EMS}(t) + P_2^{SS_1}(t)P_2^{EMS}(t) + P_3^{SS_1}(t)P_1^{EMS}(t) + P_3^{SS_1}(t)P_2^{EMS}(t),$$

$$P_4^{SS_3}(t) = P_4^{SS_1}(t)P_1^{EMS}(t) + P_4^{SS_1}(t)P_2^{EMS}(t),$$

$$P_5^{SS_3}(t) = P_5^{SS_1}(t)P_1^{EMS}(t) + P_5^{SS_1}(t)P_2^{EMS}(t) + P_3^{EMS}(t).$$

- L_z-transforms for full conventional diesel-electric power system

$$\begin{aligned}L_z\{G^{CDEPS}(t)\} &= \Omega_{f_{ser}}\left(L_z\{G^{SS_3}(t)\}, L_z\{G^{FPP}(t)\}\right) \\ &= \Omega_{f_{ser}}\left(P_1^{SS_3}(t)z^{5500} + P_2^{SS_3}(t)z^{4125} + P_3^{SS_3}(t)z^{2750} + P_4^{SS_3}(t)z^{1375}\right. \\ &\quad \left. + P_5^{SS_3}(t)z^0, p_1^{FPP}(t)z^{5500} + p_2^{FPP}(t)z^0\right) \\ &= P_1^{CDEPS}(t)z^{5500} + P_2^{CDEPS}(t)z^{4125} + P_3^{CDEPS}(t)z^{2750} \\ &\quad + P_4^{CDEPS}(t)z^{1375} + P_5^{CDEPS}(t)z^0.\end{aligned}$$

(5.25)

where:

$$P_1^{CDEPS}(t) = P_1^{SS_3}(t)p_1^{FPP}(t),$$

$$P_2^{CDEPS}(t) = P_2^{SS_3}(t)p_1^{FPP}(t),$$

$$P_3^{CDEPS}(t) = P_3^{SS_3}(t)p_1^{FPP}(t),$$

$$P_4^{CDEPS}(t) = P_4^{SS_3}(t)p_1^{FPP}(t),$$

$$P_5^{CDEPS}(t) = P_5^{SS_3}(t)p_1^{FPP}(t) + p_2^{FPP}(t).$$

5.3.5 Multi-State Model for Ship's Alternative Diesel-Electric Traction Drive

As was shown in Figure 5.3, the multi-state model for the multi-power source traction drive may be presented as two similar subsystems connected in parallel with fixed pitch propellers connected in series. Each subsystem consists of a diesel-generator subsystem, a main switchboard, an electrical energy converter, and an electric motor. Therefore, the L_z-transform of the whole system is as follows:

$$L_z\{G^{ADEPS}(t)\} = \Omega_{f_{ser}}$$

$$\left[\Omega_{f_{par}}\left\{\Omega_{f_{ser}}\left(L_z\{G^{DGS_1}(t)\}, L_z\{G^{MSb_1}(t)\}, L_z\{G^{EEC_1}(t)\}, L_z\{G^{EM_1}(t)\}\right)\right.\right.$$

$$\left.\Omega_{f_{ser}}\left(L_z\{G^{DGS_2}(t)\}, L_z\{G^{MSb_2}(t)\}, L_z\{G^{EEC_2}(t)\}, L_z\{G^{EM_2}(t)\}\right)\right\}, \quad (5.26)$$

$$\left. L_z\{G^{FPP}(t)\}\right]$$

Using the recursive derivation approach [5], we will calculate the L_z-transform of the whole system as follows:

$$L_z\{G^{SS_{i1}}(t)\} = \Omega_{f_{ser}}\left(L_z\{G^{DGS_i}(t)\}, L_z\{G^{MSb_i}(t)\}\right), \quad i = 1, 2$$

$$L_z\{G^{SS_{i2}}(t)\} = \Omega_{f_{ser}}\left(L_z\{G^{SS_{1i}}(t)\}, L_z\{G^{EEC_i}(t)\}\right), \quad i = 1, 2$$

$$L_z\{G^{SS_{i3}}(t)\} = \Omega_{f_{ser}}\left(L_z\{G^{SS_{2i}}(t)\}, L_z\{G^{EM_i}(t)\}\right), \quad i = 1, 2$$

$$L_z\{G^{SS_3}(t)\} = \Omega_{f_{par}}\left(L_z\{G^{SS_{13}}(t)\}, L_z\{G^{SS_{23}}(t)\}\right)$$

$$L_z\{G^{ADEPS}(t)\} = \Omega_{f_{ser}}\left(L_z\{G^{SS_3}(t)\}, L_z\{G^{FPP}(t)\}\right)$$

As before, the Ω_{fser} operator is used and we obtain the following L_z-transforms:

- L_z-transforms for SS_{i1} subsystem

$$L_z\{G^{SS_{i1}}(t)\} = \Omega_{f_{ser}}\left(L_z\{G^{DGS_i}(t)\}, L_z\{G^{MSb_i}(t)\}\right)$$

$$= \Omega_{f_{ser}}\left(P_1^{DGS_i}(t)z^{2750} + P_2^{DGS_i}(t)z^{1375} + P_3^{DGS_i}(t)z^0,\right. \quad (5.27)$$

$$\left. p_1^{MSb_i}(t)z^{2750} + p_2^{MSb_i}(t)z^0\right)$$

$$= P_1^{SS_{i1}}(t)z^{2750} + P_2^{SS_{i1}}(t)z^{1325} + P_3^{SS_{i1}}(t)z^0.$$

where:

$$P_1^{SS_{i1}}(t) = P_1^{DGS_i}(t) p_1^{MSb_i}(t),$$

$$P_2^{SS_{i1}}(t) = P_2^{DGS_i}(t) p_1^{MSb_i}(t),$$

$$P_3^{SS_{i1}}(t) = P_3^{DGS_i}(t) p_1^{MSb_i}(t) + p_2^{MSb_i}(t).$$

- L_z-transforms for SS_{i2} subsystem

$$\begin{aligned}
L_z\{G^{SS_{i2}}(t)\} &= \Omega_{f_{ser}}\left(L_z\{G^{SS_{i1}}(t)\}, L_z\{G^{EEC_i}(t)\}\right) \\
&= \Omega_{f_{ser}}\left(P_1^{SS_{i1}}(t) z^{2750} + P_2^{SS_{i1}}(t) z^{1325} + P_3^{SS_{i1}}(t) z^0,\right. \\
&\qquad\left. p_1^{EEC_i}(t) z^{2750} + p_2^{EEC_i}(t) z^0\right) \\
&= P_1^{SS_{i2}}(t) z^{2750} + P_2^{SS_{i2}}(t) z^{1325} + P_3^{SS_{i2}}(t) z^0.
\end{aligned}$$ (5.28)

where:

$$P_1^{SS_{i2}}(t) = P_1^{SS_{i1}}(t) p_1^{EEC_i}(t),$$

$$P_2^{SS_{i2}}(t) = P_2^{SS_{i1}}(t) p_1^{EEC_i}(t),$$

$$P_3^{SS_{i2}}(t) = P_3^{SS_{31}}(t) p_1^{EEC_i}(t) + p_2^{EEC_i}(t).$$

- L_z-transforms for SS_{i3} subsystem

$$\begin{aligned}
L_z\{G^{SS_{i3}}(t)\} &= \Omega_{f_{ser}}\left(L_z\{G^{SS_{i2}}(t)\}, L_z\{G^{EM_i}(t)\}\right) \\
&= \Omega_{f_{ser}}\left(P_1^{SS_{i2}}(t) z^{2750} + P_2^{SS_{i2}}(t) z^{1325} + P_3^{SS_{i2}}(t) z^0,\right. \\
&\qquad\left. p_1^{EM_i}(t) z^{2750} + p_2^{EM_i}(t) z^0\right) \\
&= P_1^{SS_{i3}}(t) z^{2750} + P_2^{SS_{i3}}(t) z^{1375} + P_3^{SS_{i3}}(t) z^0.
\end{aligned}$$ (5.29)

where:

$$P_1^{SS_{i3}}(t) = P_1^{SS_{i2}}(t) p_1^{EM_i}(t),$$

$$P_2^{SS_{i3}}(t) = P_2^{SS_{i2}}(t) p_1^{EM_i}(t),$$

$$P_3^{SS_{i3}}(t) = P_3^{SS_{i2}}(t) p_1^{EM_i}(t) + p_2^{EM_i}(t).$$

- L_z-transforms for SS_3 subsystem

$$L_z\left\{G^{SS_3}(t)\right\} = \Omega_{f_{par}}\left(L_z\left\{G^{SS_{13}}(t)\right\}, L_z\left\{G^{SS_{23}}(t)\right\}\right)$$

$$= \Omega_{f_{par}}\left(P_1^{SS_{13}}(t)z^{2750} + P_2^{SS_{13}}(t)z^{1375} + P_3^{SS_{13}}(t)z^0,\right. \tag{5.30}$$

$$\left. P_1^{SS_{23}}(t)z^{2750} + P_2^{SS_{23}}(t)z^{1375} + P_3^{SS_{23}}(t)z^0\right)$$

$$= P_1^{SS_3}(t)z^{5500} + P_2^{SS_3}(t)z^{4125} + P_3^{SS_3}(t)z^{2750} + P_4^{SS_3}(t)z^{1375} + P_5^{SS_3}(t)z^0.$$

where:

$$P_1^{SS_3}(t) = P_1^{SS_{13}}(t)P_1^{SS_{23}}(t),$$

$$P_2^{SS_3}(t) = P_1^{SS_{13}}(t)P_2^{SS_{23}}(t) + P_2^{SS_{13}}(t)P_1^{SS_{23}}(t),$$

$$P_3^{SS_3}(t) = P_1^{SS_{13}}(t)P_3^{SS_{23}}(t) + P_2^{SS_{13}}(t)P_2^{SS_{23}}(t) + P_3^{SS_{13}}(t)P_1^{SS_{23}}(t),$$

$$P_4^{SS_3}(t) = P_2^{SS_{13}}(t)P_3^{SS_{23}}(t) + P_3^{SS_{13}}(t)P_2^{SS_{23}}(t),$$

$$P_5^{SS_3}(t) = P_3^{SS_{13}}(t)P_3^{SS_{23}}(t).$$

- L_z-transforms for full alternative diesel-electric power system

$$L_z\left\{G^{ADEPS}(t)\right\} = \Omega_{f_{ser}}\left(L_z\left\{G^{SS_3}(t)\right\}, L_z\left\{G^{FPP}(t)\right\}\right)$$

$$= \Omega_{f_{ser}}\left(P_1^{SS_3}(t)z^{5500} + P_2^{SS_3}(t)z^{4125} + P_3^{SS_3}(t)z^{2750}\right.$$

$$\left. + P_4^{SS_3}(t)z^{1375} + P_5^{SS_3}(t)z^0, p_1^{FPP}(t)z^{5500} + p_2^{FPP}(t)z^0\right) \tag{5.31}$$

$$= P_1^{ADEPS}(t)z^{5500} + P_2^{ADEPS}(t)z^{4125} + P_3^{ADEPS}(t)z^{2750}$$

$$+ P_4^{ADEPS}(t)z^{1375} + P_5^{ADEPS}(t)z^0.$$

where:

$$P_1^{ADEPS}(t) = P_1^{SS_3}(t)p_1^{FPP}(t),$$

$$P_2^{ADEPS}(t) = P_2^{SS_3}(t)p_1^{FPP}(t),$$

$$P_3^{ADEPS}(t) = P_3^{SS_3}(t)p_1^{FPP}(t),$$

$$P_4^{ADEPS}(t) = P_4^{SS_3}(t)p_1^{FPP}(t),$$

$$P_5^{ADEPS}(t) = P_5^{SS_3}(t)p_1^{FPP}(t) + p_2^{FPP}(t).$$

5.3.6 Calculation Reliability Indices of Multi-Power Source Traction Drives

From Equations 5.3, 5.25, and 5.31, we obtain the MSS instantaneous availability when the demand is constant and equal to w:

- Winter period: 75% demand level

$$A_{w \geq 4125KW}^{CDEPS}(t) = \sum_{g_i^{CDEPS} \geq 4125} P_i^{CDEPS}(t) = P_1^{CDEPS}(t) + P_2^{CDEPS}(t),$$

$$A_{w \geq 4125KW}^{ADEPS}(t) = \sum_{g_i^{ADEPS} \geq 4125} P_i^{ADEPS}(t) = P_1^{ADEPS}(t) + P_2^{ADEPS}(t). \quad (5.32)$$

- Summer period: 50% demand level

$$A_{w \geq 2750KW}^{CDEPS}(t) = \sum_{g_i^{CDEPS} \geq 2750} P_i^{CDEPS}(t)$$

$$= P_1^{CDEPS}(t) + P_2^{CDEPS}(t) + P_2^{CDEPS}(t),$$

$$A_{w \geq 2750KW}^{ADEPS}(t) = \sum_{g_i^{ADEPS} \geq 2750} P_i^{ADEPS}(t)$$

$$= P_1^{ADEPS}(t) + P_2^{ADEPS}(t) + P_3^{ADEPS}(t). \quad (5.33)$$

Turning now to MSS instantaneous power performance, from Equations 5.4, 5.25, and 5.31 for the multi-power source traction drive, we get:

$$E^{CDEPS}(t) = \sum_{g_i^{DEPS} > 0} g_i^{CDEPS} P_i^{CDEPS}(t) = \sum_{i=1}^{4} g_i^{CDEPS} P_i^{CDEPS}(t)$$

$$= 5500 \cdot P_1^{CDEPS}(t) + 4125 \cdot P_2^{CDEPS}(t) + 2750 \cdot P_3^{CDEPS}(t) + 1375 \cdot P_4^{CDEPS}(t),$$

$$E^{ADEPS}(t) = \sum_{g_i^{DEPS} > 0} g_i^{ADEPS} P_i^{ADEPS}(t) = \sum_{i=1}^{4} g_i^{ADEPS} P_i^{ADEPS}(t)$$

$$= 5500 \cdot P_1^{ADEPS}(t) + 4125 \cdot P_2^{ADEPS}(t) + 2750 \cdot P_3^{ADEPS}(t) + 1375 \cdot P_4^{ADEPS}(t). \quad (5.34)$$

MSS instantaneous power deficiency of the multi-power source traction drive for the winter period can be obtained as follows from Equations 5.5, 5.25, and 5.31:

$$D_{w\geq 4125KW}^{CDEPS}(t) = \sum_{i=3}^{5} P_i^{CDEPS}(t) \cdot \max(4125 - g_i, 0)$$

$$= 1375 \cdot P_3^{CDEPS}(t) + 2750 \cdot P_4^{CDEPS}(t) + 4125 \cdot P_5^{CDEPS}(t), \qquad (5.35)$$

$$D_{w\geq 4125KW}^{ADEPS}(t) = \sum_{i=3}^{5} P_i^{ADEPS}(t) \cdot \max(4125 - g_i, 0)$$

$$= 1375 \cdot P_3^{ADEPS}(t) + 2750 \cdot P_4^{ADEPS}(t) + 4125 \cdot P_5^{ADEPS}(t).$$

MSS instantaneous power deficiency of the multi-power source traction drive for the summer period is given by:

$$D_{w\geq 2750KW}^{CDEPS}(t) = \sum_{i=4}^{5} P_i^{CDEPS}(t) \cdot \max(2750 - g_i, 0)$$

$$= 1375 \cdot P_4^{CDEPS}(t) + 2750 \cdot P_5^{CDEPS}(t), \qquad (5.36)$$

$$D_{w\geq 2750KW}^{ADEPS}(t) = \sum_{i=4}^{5} P_i^{ADEPS}(t) \cdot \max(2750 - g_i, 0)$$

$$= 1375 \cdot P_4^{ADEPS}(t) + 2750 \cdot P_5^{ADEPS}(t).$$

Table 5.1 provides failure and repair rates (in years^{-1}) of each system's elements.

Figure 5.5 displays the behavior of the MSS mean instantaneous availability for various fixed demand levels. By visual inspection of Figure 5.5, we conclude that the instantaneous availability for the 75% demand level is greater for the conventional schema, and for the summer period it is greater for the alternative schema.

TABLE 5.1

Failure and Repair Rates of Systems Elements (in Years^{-1})

	Failure Rates	Repair Rates
Diesel engine	2.2	73
Generator	0.15	175.2
Main switchboard	0.05	584
Electrical energy converter	0.2	673
Electric motor	0.26	117
Fixed pitch propeller	0.01	125

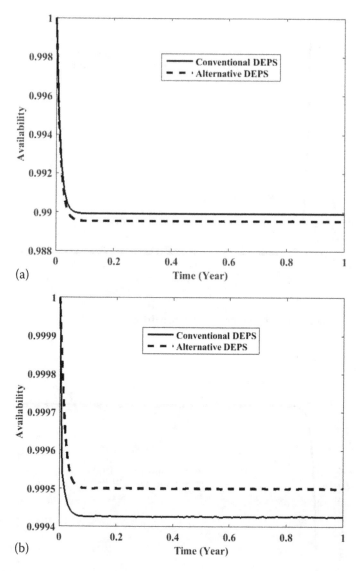

FIGURE 5.5
(a) MSS instantaneous availability for winter period (75% demand level). (b) MSS instantaneous availability for summer period (50% demand level).

The power performance for the multi-power source traction drive (Figure 5.6) for both schemas is the same and, after 10 days, exploitation becomes 5320 kW.

The power performance deficiency for the multi-power source traction drive (Figure 5.7) for both schemas is also the same for the winter and summer periods.

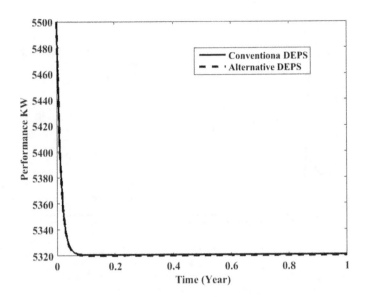

FIGURE 5.6
Power performance for multi-power source traction drive.

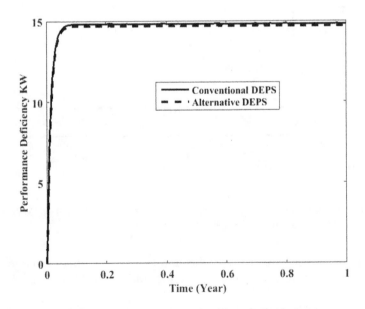

FIGURE 5.7
MSS performance deficiency for winter period (75% demand level)

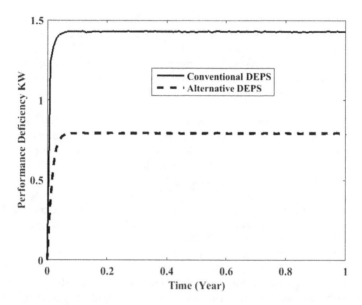

FIGURE 5.7
MSS performance deficiency for summer period (50% demand level).

5.4 Concluding Remarks

The purpose of this chapter was the implementation of the well-known L_z-transform technique to compare three important parameters of a ship's operational sustainability: availability, power performance, and power performance deficiency of the multi-state multi-power source traction drive.

The findings of this work confirm the superiority of the L_z-transform technique, which simplified the solution considerably. Note that with a Markov method, one would have required building and solving the model with 8,192 and 16,384 states, respectively.

References

1. Bolvashenkov, I. and H.-G. Herzog. Use of stochastic models for operational efficiency analysis of multi power source traction drives, in: *Proceedings of the Second International Symposium on Stochastic Models in Reliability Engineering, Life Science* and *Operations Management*, (SMRLO16), I. Frenkel and A. Lisnianski (Eds.), 15–18 February 2016, Beer Sheva, Israel, pp. 124–130.

2. Frenkel, I., I. Bolvashenkov, H.-G. Herzog, and L. Khvatskin, Performance availability assessment of combined multi power source traction drive considering real operational conditions, *Transport and Telecommunication*, 17(3), 2016, pp. 179–191.
3. Frenkel, I., I. Bolvashenkov, H.-G. Herzog, and L. Khvatskin, Operational sustainability assessment of multi power source traction drive, in: *Mathematics Applied to Engineering*, M. Ram and J.P. Davim (Eds.), Elsevier, London, pp. 191–203.
4. Jia, H., W. Jin, Y. Ding, Y. Song, and D. Yu. 2017. Multi-state time-varying reliability evaluation of smart grid with flexible demand resources utilizing Lz transform, in: *Proceedings of the International Conference on Energy Engineering and Environmental Protection (EEEP2016)*, IOP Publishing, IOP Conf. Series: Earth and Environmental Science 52 (2017).
5. Lisnianski, A., I. Frenkel, and Y. Ding, *Multi-State System Reliability Analysis and Optimization for Engineers and Industrial Managers*, Springer, London, 2010.
6. Trivedi, K., *Probability and Statistics with Reliability, Queuing and Computer Science Applications*, Wiley, New York, 2002.
7. Ushakov, I., A universal generating function, *Soviet Journal of Computer and System Sciences*, 24, 1986, pp. 37–49.
8. Yu, H., J. Yang, and H. Mo, Reliability analysis of repairable multi-state system with common bus performance sharing, *Reliability Engineering and System Safety*, 132, 2014, pp. 90–96.

6

Reliability Indicators for Hidden Markov Renewal Models

Irene Votsi

CONTENTS

6.1 Introduction 105
6.2 Hidden Markov Renewal Models 107
6.3 Conditional Failure Occurrence Rates 111
    6.3.1 Full Conditional ROCOF 111
    6.3.2 Right Conditional ROCOF 113
    6.3.3 Left Conditional ROCOF 116
6.4 Concluding Remarks 118
References 119

6.1 Introduction

Hidden Markov models (HMMs) are stochastic models that are widely used in numerous applications in reliability and DNA analysis [1], seismology [2], speech recognition [3], and so forth. They are mostly used to describe systems that are observed at discrete times. However, the observations are induced by an underlying (*hidden*) process that is unknown or unexplained. In the simplest case, this underlying process is a first-order, homogeneous Markov chain whose state space is finite.

Since the underlying process is a Markov chain, the sojourn times (i.e., the times between the successive visited states) are considered to follow the geometric distribution. However, in practice there is no clear evidence that the sojourn times should follow this particular distribution. In other words, there is no evidence that favors the choice of a Markov chain as the underlying chain over the most general semi-Markov chain. If the underlying process is a semi-Markov chain, then the corresponding model is a hidden semi-Markov model (HSMM) and, therefore, HSMMs constitute an important extension of HMMs [4–7].

In contrast to HMMs, HSMMs enable us to describe systems whose hidden states evolve based on their last visited state (Markov property) and on the time elapsed since its last visit. As a result, popular *memory-full* distributions, such as the Weibull or the shifted Poisson distributions, could be employed to describe the sojourn times. HSMMs were introduced in [8] for an application in speech recognition and, since then, they have been extensively studied and applied in many scientific disciplines. For the nonparametric HSMM, the maximum likelihood estimators (MLEs) and their asymptotic behavior were studied in [9]. Later on, the asymptotic properties of the MLEs for HSMMs defined in a general state space were studied in [10].

In the literature of semi-Markov models (SMMs), many reliability indicators have been introduced, including availability, maintainability, mean times to failure, hazard rates, and so on. For recent advances in the topic concerning discrete time SMMs, see [1, 11–13]. For continuous-time SMMs, we address the interested reader to [14] and [15]. For advances in estimation methods of nonparametric SMMs, see [10] and the references therein.

Here, we concentrate on the failure occurrence rate (ROCOF), which is a fundamental reliability indicator for random systems that could experience multiple failures. In a continuous-time context, the value of ROCOF at time $t \in \mathbb{R}^*$ represents the derivative of the mean number of failures that occurred up to time t, whereas in a discrete time context ($k \in \mathbb{N}^*$) it represents the probability that a failure occurs at time k. The failure occurrence rate may be increasing or decreasing, which means that the system of interest may improve or degenerate, respectively. On the other hand, a constant ROCOF means that the state of the system does not change over time.

ROCOF was first introduced for first-order Markov processes defined in a finite state space in [16]. More recently, the previous results were generalized for higher-order Markov processes [17]. In a continuous-time semi-Markov context, ROCOF was studied in [18] for finite state space processes and in [19] for general state space processes, respectively. The discrete time counterpart of ROCOF was evaluated for semi-Markov chains in [20]. The authors proposed an empirical estimator and applied their results to seismological data. In particular, they estimated the time-varying probability of the occurrence of a strong earthquake in Aegean Sea. Later on, ROCOF was studied for discrete hidden Markov renewal chains (HMRCs) in [21] for the case where failures are associated with specific observations rather than hidden states. Counting processes were used to define an empirical estimator whose asymptotic behavior was studied. In particular, the estimator was shown to be consistent and its asymptotic normal distribution was obtained by means of the delta method.

In the previous studies, ROCOF was defined as a *global* reliability indicator, in the sense that it was defined by means of all the starting functioning states and all the ending inoperative states. In this sense, it does not distinguish between the starting functioning state or the ending inoperative state. It takes into account all the starting operational and all the ending defective states simultaneously. However, the study of the influence of each state of the system (functioning or unworkable) in the evaluation of ROCOF could be of special interest. To the best of our knowledge, no studies have been carried out on the impact of the starting functioning states and/or the ending unworkable states on the ROCOF. To address this shortcoming in the context of HMRCs, we introduce the conditional counterparts of the failure occurrence rate. The study of these elementary rates may highlight the functioning states that mostly influence the ROCOF and give insight into the dynamics of failures. In this sense, these indicators may enable us to identify opportunities for improving reliability performance, which is of special interest for real-life applications. We further present consistent and asymptotically normal empirical estimators of the conditional versions of ROCOF.

The chapter is organized as follows. In Section 6.2, we present the notation and preliminaries of HMRCs. Section 6.3 describes the definition, evaluation, and statistical estimation of the conditional rates of failure occurrences. Finally, in Section 6.4, we give some concluding remarks.

6.2 Hidden Markov Renewal Models

First, we denote by $\mathbf{J} = (J_n)_{n \in \mathbb{N}}$ a Markov chain (*embedded Markov chain [EMC]*) with finite state space $E = \{1, \ldots, s\}$ and by $\mathbf{S} = (S_n)_{n \in \mathbb{N}}$ a sequence of jump times, that is, times when the EMC visits different successive states. We further denote by $\mathbf{X} = (X_n)_{n \in \mathbb{N}}$ the sequence of sojourn (or inter-arrival) times, where $X_0 = S_0 = 0$ almost surely, and $X_n = S_n - S_{n-1}$ for all $n \in \mathbb{N}^*$. The semi-Markov chain $\mathbf{Z} = (Z_k)_{k \in \mathbb{N}}$ is defined by $Z_k = J_{N(k)}$, where $N(k) = \max\{n \geq 0 : S_n \leq k\}$, $k \in \mathbb{N}$. The stochastic process $(\mathbf{J}, \mathbf{S}) = (J_n, S_n)_{n \in \mathbb{N}}$, called a *Markov renewal chain (MRC)*, is considered to be (time) homogeneous and is characterized by the semi-Markov kernel (SMK) $\mathbf{q} = (q_{ij}(k); i, j \in E, k \in \mathbb{N})$, where

$$q_{ij}(k) = P(J_{n+1} = j, X_{n+1} = k \mid J_n = i)$$

for any $i, j \in E$ and $k, n \in \mathbb{N}$.

The evaluation of the conditional ROCOF requires, additionally, the definition of the ℓ-fold convolution of the SMK $\mathbf{q}^{(\ell)} = (q_{ij}^{(\ell)}(k); i, j \in E, k \in \mathbb{N})$, where

$$q_{ij}^{(\ell)}(k) = P(J_\ell = j, S_\ell = k \mid J_0 = i),$$

for any $i, j \in E$ and $\ell, k \in \mathbb{N}$. The ℓ-fold convolution of the semi-Markov kernel may further be obtained recursively by means of the lower-order convolutions of the semi-Markov kernel [1].

We further denote by $\mathbf{a} = (a(i); i \in E)$, where $a(i) = P(J_0 = i)$, the initial law of the EMC. The transition probability matrix of the EMC is defined by

$$\mathbf{P} = (p_{ij}; i, j \in E)$$

where:

$p_{ij} = P(J_{n+1} = j | J_n = i)$

$i, j \in E$,

$n \in \mathbb{N}$

and the conditional distribution of the sojourn time in state $i \in E$, given that the next visited state is $j \in E$, is given by $f_{ij}(k) = P(X_{n+1} = k | J_n = i, J_{n+1} = j)$, $k \in \mathbb{N}$. Here, we assume that there are neither instantaneous transitions nor self-transitions of the EMC. Moreover, we define the survival function of the sojourn time in state $i \in E$ by $\bar{H}_i(k) = P(X_{n+1} > k | J_n = i)$, $k \in \mathbb{N}$ and the corresponding mean sojourn time by $m_i = \mathbb{E}(S_1 | J_0 = i)$.

In what follows, we consider that the following assumptions are fulfilled:

A1. The EMC **J** is ergodic.
A2. For any state $i \in E$, $m_i = \mathbb{E}(S_1 | J_0 = i)$ is finite.
A3. The MRC (**J**, **S**) is aperiodic.

Second, we denote by $\mathbf{Y} = (Y_n)_{n \in \mathbb{N}}$ an observation sequence that is defined in a finite state space A and is recorded only at jump times $n \in \mathbb{N}$. We assume that the observations are conditionally independent given the underlying states of the EMC, that is,

$$P(Y_n = a | Y_0, \ldots, Y_{n-1}, J_0, \ldots, J_n = i) = P(Y_n = a | J_n = i),$$

for any $a \in A, i \in E, n \in \mathbb{N}$.

Here, we consider that observations are recorded at jump times and call the process $(\mathbf{J}, \mathbf{S}, \mathbf{Y}) = (J_n, S_n, Y_n)_{n \in \mathbb{N}}$ a *hidden Markov renewal chain* (Figure 6.1). The state space of $(\mathbf{J}, \mathbf{S}, \mathbf{Y})$ is denoted by E^* ($|E^*| = d$) and its initial law by $\boldsymbol{\alpha} = (\alpha(i, y); i \in E, y \in A)$, where $\alpha(i, y) = P(J_0 = i, S_0 = 0, Y_0 = y)$. In contrast, if we consider that observations are recorded at calendar times $k \in \mathbb{N}$ (and not just at jump times $n \in \mathbb{N}$), then the process $(\mathbf{Z}, \mathbf{Y}) = (Z_k, Y_k)_{k \in \mathbb{N}}$ is a *hidden semi-Markov chain* [1].

We consider that A is divided into two disjoint, non-empty subsets U and D such that $U, D \subset A$ ($U, D \neq A$). The subset $U = \{1, \ldots, r\}$ represents the functioning (or up) states, whereas the subset $D = \{r+1, \ldots, s\}$ stands for the unworkable (or down) states. We denote by $L = E \times \mathbb{N}$ ($|L| = \ell$) the state space of the MRC and by $\boldsymbol{\pi} = (\pi(i, s); i \in E, s \in \mathbb{N})$, where $\pi(i, s) = \sum_{u \in E} v(u) q_{ui}(s)$, its stationary distribution. Since we assume stationarity, the stationary distribution of the EMC, $\mathbf{v} = (v(i); i \in E)$ coincides with its initial distribution, **a**. We further define $R_i(U) = P(Y_n \in U | J_n = i)$ and $R_i(D) = P(Y_n \in D | J_n = i)$, for any $i \in E, n \in \mathbb{N}$.

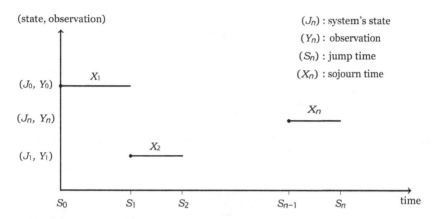

FIGURE 6.1
A representative sample path of the process $(J_n, S_n, Y_n)_{n \in \mathbb{N}}$.

We now turn our attention to statistical estimation aspects with regards to HMRCs. We first consider a sample path of the HMRC up to fixed arbitrary time $M \in \mathbb{N}$, that is,

$$\mathcal{H}(M) = \left(J_0, Y_0, S_1, \ldots, J_{N(M)}, Y_{N(M)}, S_{N(M)}, U_M \right),$$

where $U_M = M - S_{N(M)}$. Then we denote $S_0^M = (S_0, \ldots, S_{N(M)})$, $Y_0^M = (Y_0, \ldots, Y_{N(M)})$, $J_0^M = (J_0, \ldots, J_{N(M)})$ and $T_M = \{0, \ldots, M\}$. We further denote by δ Kronecker's delta and define the (block diagonal) covariance matrix Δ, of dimensions $(\ell d) \times (\ell d)$, by

$$\Delta = \begin{pmatrix} \Phi(1,0) & 0 & \cdots & & \cdots & 0 \\ \vdots & \ddots & & \vdots & & \vdots \\ \vdots & & \Phi(i,t) & \vdots & & \vdots \\ \vdots & & \vdots & \ddots & & \vdots \\ 0 & \cdots & \cdots & & \cdots & \Phi(s, S_{N(M)}) \end{pmatrix}$$

where

$$\Phi(i,t) = \frac{1}{\pi(i,t)} \left(\delta_{(j,t',m)(r,t'',q)} R_{j;m} q_{ij}(t'-t) - q_{ij}(t'-t) R_{j;m} R_{r;q} q_{ir}(t''-t) \right)$$

for any $(i,t) \in L$, $(j,t',m), (r,t'',q) \in E^*$.

Second, we introduce the counting processes of transitions of $(J_n)_{n \in \mathbb{N}}$ up to time M

$$N_{ij}(M) = \sum_{n=1}^{N(M)} 1_{\{J_{n-1}=i, J_n=j\}},$$

$$N_{ij}(k,M) = \sum_{n=1}^{N(M)} 1_{\{J_{n-1}=i, X_n=k, J_n=j\}},$$

where $i, j \in E$, $k \in \mathbb{N}^*$. We further introduce the counting process of visits of $(J_n)_{n \in \mathbb{N}}$ and $(Y_n)_{n \in \mathbb{N}}$ up to time M:

$$N_{i;a}(M) = \sum_{n=1}^{N(M)} 1_{\{J_{n-1}=i, Y_{n-1}=a\}},$$

where $i \in E$, $a \in A$.

Since we assume stationarity, the parameter set of the model is $\vartheta = (\mathbf{q}, \mathbf{R})$. Moreover, the complete likelihood function, that is, the likelihood function that is based on the complete data, is denoted by $f_M(J_0^M, Y_0^M, S_0^M \mid \vartheta)$. Then, the partial likelihood function, which is based on incomplete data, is given by

$$g_M(Y_0^M \mid \vartheta) = \sum_{J_0^M \in E^{N(M)+1}} \sum_{S_0^M \in T_M^{N(M)}} f_M(J_0^M, Y_0^M, S_0^M \mid \vartheta).$$

The complete log-likelihood function is expressed in terms of the aforementioned counting processes, as follows:

$$\log f_M(J_0^M, Y_0^M, S_0^M \mid \vartheta) = \log \left(\prod_{i,j \in E} p_{ij}^{N_{ij}(M)} \prod_{i,j \in E} \prod_{k \in \mathbb{N}^*} f_{ij}^{N_{ij}(k,M)} \prod_{i \in E} \prod_{a \in A} R_{i;a}^{N_{i;a}(M)} \overline{H}_i(k) \right)$$

$$= \sum_{i,j \in E} N_{ij}(M) \log p_{ij} + \sum_{i,j \in E} \sum_{k \in \mathbb{N}^*} N_{ij}(k,M) \log f_{ij}(k)$$

$$+ \sum_{i \in E} \sum_{a \in A} N_{i;a}(M) \log R_{i;a} + \log \left(\sum_{k \geq M - S_{N(M)}} \overline{H}_i(k) \right).$$

The MLEs of the parameters

$$\vartheta_{\mathrm{MLE}} = \arg\max_{\vartheta} g_M(Y_0^M \mid \vartheta),$$

may be obtained by an adaptation of the expectation maximization (EM) algorithm [22] (as presented in [1]) in the case where observations are recorded at jump times. In the following, we denote the MLEs of $q_{ij}(k)$, $q_{ij}^{(\ell)}(k)$,

$R_{i;a}$, and $R_i(A_1)$ by $\hat{q}_{ij}(k,M)$, $\hat{q}_{ij}^{(\ell)}(k,M)$, $\hat{R}_{i;a}(M)$, and $\hat{R}_i(A_1,M)$, respectively, for any $i,j \in E, k, \ell \in \mathbb{N}, a \in A, A_1 \subset A$.

6.3 Conditional Failure Occurrence Rates

Our main objective is to study reliability indicators that could be used to improve the performance of random, repairable systems that are encountered in real-life problems. These indicators could be taken into account in the development of maintenance strategies and in decision making. In particular, we focus on the conditional failure occurrence rates.

6.3.1 Full Conditional ROCOF

First of all, we are interested in studying the impact of both the starting up and the ending down states on the ROCOF, simultaneously. To do that, we define the full conditional ROCOF as the mean transition number of the HMRC to the set D at time k, starting in state $i \in U$ and ending in state $j \in D$, that is,

$$r_{ij}^{\#}(k) = \mathbb{E}\left(N_{ij}^{\#}(k)\right) - \mathbb{E}\left(N_{ij}^{\#}(k-1)\right),$$

where

$$N_{ij}^{\#}(k) = \sum_{l=1}^{k} \mathbf{1}_{\{Y_{l-1}=i, Y_l=j\}}.$$

Proposition 1. The full conditional ROCOF at time $k \in \mathbb{N}^*$ is given by

$$r_{ij}^{\#}(k) = \sum_{(\ell,y) \in E \times A} \sum_{l=1}^{k} \sum_{s_0, s_1 \in E} \sum_{k_0=0}^{k} a(\ell, y) R_{\ell;y} R_{s_0;i} R_{s_1;j} q_{is_0}^{(l-1)}(k_0) q_{s_0 s_1}(k-k_0),$$

for any fixed, arbitrary states $i \in U, j \in D$.

Proof 1. For fixed $k \in \mathbb{N}^*, i \in U, j \in D$ we define

$$r_{(\ell,y,i,j)}^{\#}(k) = \mathbb{E}\left(N_{ij}^{\#}(k) - N_{ij}^{\#}(k-1) \mid J_0 = \ell, Y_0 = y\right)$$

$$= \sum_{l=1}^{k} P\left(Y_{l-1} = i, Y_l = j, S_l = k \mid J_0 = \ell, Y_0 = y\right). \tag{6.1}$$

Moreover,

$$P(Y_{l-1} = i, Y_l = j, S_l = k \mid J_0 = \ell, Y_0 = y)$$

$$= \sum_{s_0, s_1 \in E} \sum_{k_0=0}^{k} R_{s_0;i} R_{s_1;j} q_{iso}^{(l-1)}(k_0) q_{s_0 s_1}(k - k_0). \qquad (6.2)$$

Consequently, from Equations 6.1 and 6.2, we get

$$r_{ij}^{\#}(k) = \sum_{(\ell,y) \in E \times A} a(\ell, y) R_{\ell;y} r_{(\ell,y,i,j)}^{\#}(k)$$

$$= \sum_{(\ell,y) \in E \times A} \sum_{l=1}^{k} \sum_{s_0,s_1 \in E} \sum_{k_0=0}^{k} a(\ell,y) R_{\ell;y} R_{s_0;i} R_{s_1;j} q_{iso}^{(l-1)}(k_0) q_{s_0 s_1}(k - k_0).$$

For any fixed, arbitrary states $i \in U$, $j \in D$, the full conditional ROCOF at time $k \in \mathbb{N}^*$ may be estimated by the following plug-in type estimator:

$$\hat{r}_{ij}^{\#}(k, M)$$

$$= \sum_{(\ell,y) \in E \times A} \sum_{l=1}^{k} \sum_{s_0,s_1 \in E} \sum_{k_0=0}^{k} \hat{a}(\ell,y) \hat{R}_{\ell;y}(M) \hat{R}_{s_0;i}(M) \hat{R}_{s_1;j}(M) \hat{q}_{iso}^{(l-1)}(k_0, M) \hat{q}_{s_0 s_1}(k - k_0, M).$$

Proposition 2. For any $i \in U$, $j \in D$, the estimator of the full conditional ROCOF at time $k \in \mathbb{N}$ is strongly consistent, that is,

$$\lim_{M \to \infty} \widehat{r_{ij}^{\#}}(k, M) = r_{ij}^{\#}(k)$$

with probability 1.

Proof 2. Following [1], the estimators $\hat{q}_{ij}(k, M)$, $\hat{q}_{ij}^{(n)}(k, M)$ and $\hat{R}_{i;a}(M)$ are strongly consistent. Moreover, since the estimator $\widehat{r_{ij}^{\#}}(k, M)$ includes a finite number of terms, the result is straightforward.

Proposition 3. For any $k \in \mathbb{N}^*$, $i \in U$, $j \in D$, the random vector $\sqrt{M}\left(\widehat{r_{ij}^{\#}}(k,M) - r_{ij}^{\#}(k)\right)$ is asymptotically normal, that is,

$$\sqrt{M}\left(\widehat{r_{ij}^{\#}}(k,M) - r_{ij}^{\#}(k)\right) \xrightarrow[M \to \infty]{\mathcal{L}} \mathcal{N}\left(0, \Psi_1' \Delta \Psi_1'^T\right),$$

where $\Psi_1 : \mathbb{R}^{\ell d} \to \mathbb{R}^+$ is defined by

$$\Psi_1\left(\left(R_{j';m'}, q_{i'j'}(k'-k_0')\right); (i',k_0') \in L, (i',j',k') \in E^*\right)$$

$$= \sum_{(\ell,y)\in E\times A} \sum_{l=1}^{k} \sum_{s_0,s_1\in E} \sum_{k_0=0}^{k} a(\ell,y) R_{\ell;y} R_{s_0;i} R_{s_1;j} q_{is_0}^{(l-1)}(k_0) q_{s_0 s_1}(k-k_0),$$

and

$$\Psi_1'^T = \left(\left(\frac{\partial \Psi_1}{\partial R_{j';m'}}, \frac{\partial \Psi_1}{\partial q_{i'j'}(k'-k_0')}\right); (i',k_0') \in L, (i',j',k') \in E^*\right),$$

is the column vector of derivatives of $r_{ij}^{\#}(k)$ with respect to $R_{j';m'}$ and $q_{i'j'}(k'-k_0')$.

Proof 3. First, we notice that

$$\sqrt{M}\left(\hat{r}_{ij}^{\#}(k,M) - r_{ij}^{\#}(k)\right)$$

$$= \sqrt{M} \sum_{(\ell,y)\in E\times A} \sum_{l=1}^{k} \sum_{s_0,s_1\in E} \sum_{k_0=0}^{k} \left(\hat{a}(\ell,y)\hat{R}_{\ell;y}(M)\hat{R}_{s_0;i}(M)\hat{R}_{s_1;j}(M)\hat{q}_{is_0}^{(l-1)}(k_0,M)\hat{q}_{s_0 s_1}(k-k_0,M)\right.$$

$$\left. - a(\ell,y) R_{\ell;y} R_{s_0;i} R_{s_1;j} q_{is_0}^{(l-1)}(k_0) q_{s_0 s_1}(k-k_0)\right).$$

Then, since

$$\sqrt{M}\left(\hat{q}_{i'j'}(k'-k_0',M)\hat{R}_{j';m'}(M) - q_{i'j'}(k'-k_0')R_{j';m'}\right) \xrightarrow[M\to\infty]{\mathcal{L}} \mathcal{N}(0,\Delta),$$

the delta method [23] leads directly to

$$\sqrt{M}\left(\Psi_1\left(\hat{q}_{i'j'}(k'-k_0',M),\hat{R}_{j';m'}(M)\right) - \Psi_1\left(\hat{q}_{i'j'}(k'-k_0',M),\hat{R}_{j';m'}(M)\right)\right)$$

$$\xrightarrow[M\to\infty]{\mathcal{L}} \mathcal{N}\left(0, \Psi_1'\Delta\Psi_1'^T\right).$$

6.3.2 Right Conditional ROCOF

We go one step further and aim at studying the impact of the ending down state on the ROCOF. To do that, we define the right conditional ROCOF as the mean transition number of the HMRC to the set D at time k, given that it ends in state $j \in D$, that is,

$$\bar{r}_j(k) = \mathbb{E}\left(\bar{N}_j(k)\right) - \mathbb{E}\left(\bar{N}_j(k-1)\right),$$

where

$$\bar{N}_j(k) = \sum_{l=1}^{k} 1_{\{Y_{l-1} \in U, Y_l = j\}}.$$

Proposition 4. The right conditional ROCOF at time $k \in \mathbb{N}^*$ is defined by

$$\bar{r}_j(k) = \sum_{(\ell,y) \in E \times A} \sum_{l=1}^{k} \sum_{s_0, s_1 \in E} \sum_{k_0=0}^{k} a(\ell,y) R_{\ell;y} R_{s_0}(U) R_{s_1;j} q_{is_0}^{(l-1)}(k_0) q_{s_0 s_1}(k-k_0),$$

for any fixed, arbitrary state $j \in D$.

Proof 4. For fixed $k \in \mathbb{N}^*$, $j \in D$ we define

$$\bar{r}_{(\ell,y,j)}(k) = \mathbb{E}\left(\bar{N}_j(k) - \bar{N}_j(k-1) \mid J_0 = \ell, Y_0 = y\right)$$

$$= \sum_{l=1}^{k} P(Y_{l-1} \in U, Y_l = j, S_l = k \mid J_0 = \ell, Y_0 = y). \quad (6.3)$$

Moreover,

$$P(Y_{l-1} \in U, Y_l = j, S_l = k \mid J_0 = \ell, Y_0 = y)$$

$$= \sum_{s_0, s_1 \in E} \sum_{k_0=0}^{k} R_{s_0}(U) R_{s_1;j} q_{is_0}^{(l-1)}(k_0) q_{s_0 s_1}(k-k_0). \quad (6.4)$$

Consequently, from Equations 6.3 and 6.4, we get

$$\bar{r}_j(k) = \sum_{(\ell,y) \in E \times A} a(\ell,y) R_{\ell;y} \bar{r}_{(\ell,y,j)}(k)$$

$$= \sum_{(\ell,y) \in E \times A} \sum_{l=1}^{k} \sum_{s_0, s_1 \in E} \sum_{k_0=0}^{k} a(\ell,y) R_{\ell;y} R_{s_0}(U) R_{s_1;j} q_{is_0}^{(l-1)}(k_0) q_{s_0 s_1}(k-k_0).$$

For any fixed, arbitrary state $j \in D$, the right conditional ROCOF at time $k \in \mathbb{N}^*$ may be estimated as follows:

$$\hat{\bar{r}}_j(k,M)$$

$$= \sum_{(\ell,y) \in E \times A} \sum_{l=1}^{k} \sum_{s_0, s_1 \in E} \sum_{k_0=0}^{k} \hat{a}(\ell,y) \hat{R}_{\ell;y}(M) \hat{R}_{s_0}(U,M) \hat{R}_{s_1;j}(M) \hat{q}_{is_0}^{(l-1)}(k_0,M) \hat{q}_{s_0 s_1}(k-k_0,M).$$

Proposition 5. For any $j \in D$, $k \in \mathbb{N}$, the estimator of the right conditional ROCOF at time k is strongly consistent, that is,

$$\lim_{M\to\infty} \hat{\bar{r}}_j(k,M) = \bar{r}_j(k)$$

with probability 1.

Proof 5. To prove the consistency, we work as in Proposition 2.

Proposition 6. For any $k \in \mathbb{N}^*$, $j \in D$, the random vector $\sqrt{M}\left(\hat{\bar{r}}_j(k,M) - \bar{r}_j(k)\right)$ is asymptotically normal, that is,

$$\sqrt{M}\left(\hat{\bar{r}}_j(k,M) - \bar{r}_j(k)\right) \xrightarrow[M\to\infty]{\mathcal{L}} \mathcal{N}\left(0, \Psi_2' \Delta \Psi_2'^T\right),$$

where $\Psi_2 : \mathbb{R}^{\ell d} \to \mathbb{R}^+$ is the function

$$\Psi_2\left(\left(R_{j';m'}, q_{i'j'}(k'-k_0')\right); (i',k_0') \in L, (i',j',k') \in E^*\right)$$

$$= \sum_{(\ell,y)\in E\times A} \sum_{l=1}^{k} \sum_{s_0,s_1\in E} \sum_{k_0=0}^{k} a(\ell,y) R_{\ell;y} R_{s_0}(U) R_{s_1;j} q_{is_0}^{(l-1)}(k_0) q_{s_0s_1}(k-k_0)$$

and

$$\Psi_2'^T = \left(\left(\frac{\partial \Psi_2}{\partial R_{j';m'}}, \frac{\partial \Psi_2}{\partial q_{i'j'}(k'-k_0')}\right); (i',k_0') \in L, (i',j',k') \in E^*\right)$$

is the column vector of derivatives of $\bar{r}_j(k)$ with respect to $R_{j';m'}$ and $q_{i'j'}(k'-k_0')$.

Proof 6. We first notice that

$$\sqrt{M}\left(\hat{\bar{r}}_j(k,M) - \bar{r}_j(k)\right)$$

$$= \sqrt{M} \sum_{(\ell,y)\in E\times A} \sum_{l=1}^{k} \sum_{s_0,s_1\in E} \sum_{k_0=0}^{k} \left(\hat{a}(\ell,y)\hat{R}_{\ell;y}(M)\hat{R}_{s_0}(U,M)\hat{R}_{s_1;j}(M)\hat{q}_{is_0}^{(l-1)}(k_0,M)\hat{q}_{s_0s_1}(k-k_0,M)\right.$$

$$\left. - a(\ell,y) R_{\ell;y} R_{s_0}(U) R_{s_1;j} q_{is_0}^{(l-1)}(k_0) q_{s_0s_1}(k-k_0)\right).$$

Then, since

$$\sqrt{M}\left(\hat{q}_{i'j'}(k'-k_0',M)\hat{R}_{j';m'}(M) - q_{i'j'}(k'-k_0')R_{j';m'}\right) \xrightarrow[M\to\infty]{\mathcal{L}} \mathcal{N}(0,\Delta),$$

we have

$$\sqrt{M}\left(\Psi_2\left(\hat{q}_{i'j'}(k'-k_0'),\hat{R}_{j';m'}\right) - \Psi_2\left(\hat{q}_{i'j'}(k'-k_0'),\hat{R}_{j';m'}\right)\right) \xrightarrow[M\to\infty]{\mathcal{L}} \mathcal{N}\left(0,\Psi_2'\Delta\Psi_2'^T\right).$$

6.3.3 Left Conditional ROCOF

The left conditional ROCOF is defined as the mean transition number of the HMRC to the set D at time k, given that it starts from the fixed state $i \in U$, that is,

$$\tilde{r}_i(k) = \mathbb{E}\big(\tilde{N}_i(k)\big) - \mathbb{E}\big(\tilde{N}_i(k-1)\big),$$

where

$$\tilde{N}_i(k) = \sum_{l=1}^{k} 1_{\{Y_{l-1}=i, Y_l \in D\}}.$$

Proposition 7. The left conditional ROCOF at time $k \in \mathbb{N}^*$ is defined by

$$\tilde{r}_i(k) = \sum_{(\ell,y) \in E \times A} \sum_{l=1}^{k} \sum_{s_0, s_1 \in E} \sum_{k_0=0}^{k} a(\ell, y) R_{\ell;y} R_{s_0;i} R_{s_1}(D) q_{is_0}^{(l-1)}(k_0) q_{s_0 s_1}(k - k_0),$$

for any fixed, arbitrary state $i \in U$.

Proof 7. For fixed $k \in \mathbb{N}^*$, $i \in U$, we define

$$\tilde{r}_{(\ell,y,i)}(k) = E\big(\tilde{N}_i(k) - \tilde{N}_i(k-1) \mid J_0 = \ell, Y_0 = y\big)$$

$$= \sum_{l=1}^{k} P\big(Y_{l-1} = i, Y_l \in D, S_l = k \mid J_0 = \ell, Y_0 = y\big). \quad (6.5)$$

Moreover,

$$P\big(Y_{l-1} = i, Y_l \in D, S_l = k \mid J_0 = \ell, Y_0 = y\big)$$

$$= \sum_{s_0, s_1 \in E} \sum_{k_0=0}^{k} R_{s_0;i} R_{s_1}(D) q_{is_0}^{(l-1)}(k_0) q_{s_0 s_1}(k - k_0). \quad (6.6)$$

Consequently, from Equations 6.5 and 6.6, we get

$$\tilde{r}_i(k) = \sum_{(\ell,y) \in E \times A} a(\ell, y) R_{\ell;y} \tilde{r}_{(\ell,y,i)}(k)$$

$$= \sum_{(\ell,y) \in E \times A} \sum_{l=1}^{k} \sum_{s_0, s_1 \in E} \sum_{k_0=0}^{k} a(\ell, y) R_{\ell;y} R_{s_0;i} R_{s_1}(D) q_{is_0}^{(l-1)}(k_0) q_{s_0 s_1}(k - k_0).$$

For any fixed, arbitrary state $i \in U$, the left conditional ROCOF at time $k \in \mathbb{N}^*$ may be estimated by the following plug-in type empirical estimator:

$\hat{\tilde{r}}_i(k, M)$

$$= \sum_{(\ell,y)\in E\times A} \sum_{l=1}^{k} \sum_{s_0,s_1\in E} \sum_{k_0=0}^{k} \hat{a}(\ell,y)\hat{R}_{\ell;y}(M)\hat{R}_{s_0;i}(M)\hat{R}_{s_1}(D,M)\hat{q}_{iso}^{(l-1)}(k_0,M)\hat{q}_{s_0s_1}(k-k_0,M).$$

Proposition 8. For any state $i \in U$ and any fixed arbitrary $k \in \mathbb{N}^*$, the estimator of the right conditional ROCOF at time k is strongly consistent in the sense that

$$\lim_{M\to\infty} \hat{\tilde{r}}_i(k,M) = \tilde{r}_i(k)$$

with probability 1.
Proof 8. To prove the consistency, we work as in Proposition 2.

Proposition 9. For any $k \geq 1$, $j \in D$, the random vector $\sqrt{M}\left(\hat{\tilde{r}}_i(k,M) - \tilde{r}_i(k)\right)$ is asymptotically normal, that is,

$$\sqrt{M}\left(\hat{\tilde{r}}_i(k,M) - \tilde{r}_i(k)\right) \xrightarrow[M\to\infty]{\mathcal{L}} \mathcal{N}\left(0, \Psi_3'\Delta\Psi_3'^T\right),$$

where $\Psi_3 : \mathbb{R}^{\ell d} \to \mathbb{R}^+$ is the function

$$\Psi_3\left((R_{j';m'}, q_{i'j'}(k'-k_0'));(i',k_0')\in L,(i',j',k')\in E^*\right)$$

$$= \sum_{(\ell,y)\in E\times A} \sum_{l=1}^{k} \sum_{s_0,s_1\in E} \sum_{k_0=0}^{k} a(\ell,y) R_{\ell;y} R_{s_0;i} R_{s_1}(D) q_{iso}^{(l-1)}(k_0) q_{s_0s_1}(k-k_0)$$

and $\Psi_3'^T = \left(\left(\frac{\partial \Psi_3}{\partial q_{i'j'}(k'-k_0')}, \frac{\partial \Psi_3}{\partial R_{j';m'}}\right); (i',k_0')\in L, (i',j',k')\in E^*\right)$ is the column vector of derivatives of $\tilde{r}_i(k)$ with respect to $R_{j';m'}$ and $q_{i'j'}(k'-k_0')$.
Proof 9. We first have

$$\sqrt{M}\left(\hat{\tilde{r}}_i(k,M) - \tilde{r}_i(k)\right)$$

$$= \sqrt{M} \sum_{(\ell,y)\in E\times A} \sum_{l=1}^{k} \sum_{s_0,s_1\in E} \sum_{k_0=0}^{k} \left(\hat{a}(\ell,y)\hat{R}_{\ell;y}(M)\hat{R}_{s_0;i}(M)\hat{R}_{s_1}(D,M)\hat{q}_{iso}^{(l-1)}(k_0,M)\hat{q}_{s_0s_1}(k-k_0,M)\right.$$

$$\left. - a(\ell,y) R_{\ell;y} R_{s_0;i} R_{s_1}(D) q_{iso}^{(l-1)}(k_0) q_{s_0s_1}(k-k_0)\right).$$

Then, since

$$\sqrt{M}\left(\hat{q}_{i'j'}(k'-k_0',M)\hat{R}_{j';m'}(M) - q_{i'j'}(k'-k_0')R_{j';m'}\right) \xrightarrow[M\to\infty]{\mathcal{L}} \mathcal{N}(0,\Delta),$$

we obtain directly the desired result.

The (unconditional) ROCOF for HMRCs was introduced in [21] as the mean transition number of the HMRC to the set D at time k, that is,

$$r(k) = \mathbb{E}\big(M(k) - M(k-1)\big),$$

where

$$M(k) = \sum_{l=1}^{k} 1_{\{Y_{i-1} \in U, Y_i \in D\}}.$$

Remark 1. The following relationship between the ROCOF and its conditional counterparts holds true:

$$r(k) = \sum_{i \in U} \sum_{j \in D} r_{ij}^{\#}(k) = \sum_{j \in D} \bar{r}_j(k) = \sum_{i \in U} \tilde{r}_i(k),$$

for any $k \in \mathbb{N}^*$.

6.4 Concluding Remarks

Failure occurrence rates are important risk indicators in the theory and applications of stochastic models. We encounter their distribution as reliability or survival functions in mechanical systems or biology, respectively. Many stochastic models have been developed to analyze data associated with failure occurrences. Often, these data are driven by hidden or unobservable mechanisms. Here, we propose hidden Markov renewal models to describe systems that could experience multiple random failures. We concentrate on ROCOF, which plays a critical role in reliability analysis, since its increasing value signals the degradation of the system under study, whereas its decreasing value indicates its improvement.

We study the sensitivity of ROCOF in the current up and ending down states by introducing its conditional counterparts. This is the first attempt to raise sensitivity issues of reliability indicators for hidden Markov renewal models. We provide plug-in type estimators and show that they have appealing asymptotic properties, including consistency and asymptotic normality. Once the estimators are shown to be asymptotically normal, the estimation of the confidence interval is straightforward. It is worth noting that the estimation of the (conditional) ROCOF necessitates the estimation of the corresponding parameters of the HMRC. The MLEs may be obtained by minor modifications of the EM algorithm.

The results obtained could be applied in different scientific disciplines including seismology and mechanics. To be more precise, let us present two possible applications of our results. In seismology, earthquake occurrences are governed by the actual level of the stress field, which is not observable. We could consider a hidden Markov renewal model where states correspond to stress levels and observations to earthquake magnitude classes. In this case, the conditional counterparts of ROCOF might describe the impact of each magnitude class in the occurrence of the magnitude class that includes stronger earthquakes. On the other hand, mechanical systems are subject to random loadings that may affect their integrity when they exceed certain thresholds. SMMs might be used to describe loading data and conditional failure occurrence rates could be determined to allow the monitoring of mechanical systems. Since the real state of such systems is unknown, the conditional failure occurrence rate could be studied in a hidden Markov renewal context. For recent advances in the applications of HSMMs in health monitoring, we address readers to [24].

From a theoretical point of view, the current results might be extended for general HMRCs. A topic of interest for future research is the study of (conditional) ROCOF for (general) HSMMs. In this case, the ROCOF for HMMs could be studied as a particular case. ROCOF could further be investigated when independent and identical copies of the process are observed, each over a fixed duration. Finally, the consideration of models with a more refined structure [25] might improve the understanding of the functioning of a system and eventually provide feedback for the management of random systems.

References

1. Barbu, V., Limnios, N. (2008), *Semi-Markov Chains and Hidden Semi-Markov Models toward Applications: Their Use in Reliability and DNA Analysis*, Lecture Notes in Statistics, Springer, New York.
2. Votsi, I., Limnios, N., Tsaklidis, G., Papadimitriou, E. (2012), Estimation of the expected number of earthquake occurrences based on semi-Markov models, *Methodol. Comput. Appl. Probab.*, 14(3), 685–703.
3. Rabiner, L.R. (1990), A tutorial on hidden Markov models and selected applications in speech recognition, *Readings in Speech Recognition*, Morgan Kaufmann, San Francisco, CA, 267–296.
4. Baum, L.E., Petrie, T. (1966), Statistical inference for probabilistic functions of finite state Markov chains, *Ann. Math. Statist.*, 37(6), 1554–1563.
5. Bickel, P.J., Ritov, Y., Rydén, T. (1998), Asymptotic normality of the maximum-likelihood estimator for general hidden Markov models, *Bernoulli*, 26(4), 1614–1635.
6. Douc, R., Matias, C. (2001), Asymptotics of the maximum likelihood estimator for general hidden Markov models, *Bernoulli*, 7(3), 381–420.
7. Rydén, T. (1995), Estimating the order of hidden Markov models, *Stat. A J. Theor. Appl. Stat.*, 26(4), 345–354.

8. Ferguson, J.D. (1980), Variable duration models for speech, *Proc. Symp. on the Application of Hidden Markov Models to Text and Speech*, 143–179.
9. Barbu, V., Limnios, N. (2006), Maximum likelihood estimation for hidden semi-Markov models, *C.R. Acad. Sci. Paris*, 342, 201–205.
10. Trevezas, S., Limnios, N. (2011), Exact MLE and asymptotic properties for nonparametric semi-Markov models, *J. Nonparametr. Stat.*, 23(3), 719–739.
11. Georgiadis, S., Limnios, N., Votsi, I. (2013), Reliability and probability of first occurred failure for discrete-time semi-Markov systems. In: Frenkel, I., Karagrigoriou, A., Lisnianski, A., Kleyner, A. (Eds). *Applied Reliability Engineering and Risk Analysis: Probabilistic Models and Statistical Inference* (chap. 12, 167–179). New York: Wiley & Sons.
12. Georgiadis, S. (2017), First hitting probabilities for semi-Markov chains and estimation, *Commun. Stat. Theory Method*, 46(5), 2435–2446.
13. Barbu, V., Karagrigoriou, A., Makrides, A. (2017), Semi-Markov modelling for multi-state systems, *Methodol. Comput. Appl. Probab.*, 19(4), 1011–1028.
14. Limnios, N., Oprişan, G. (2001), *Convergence of Probability Measures*, Wiley, New York, 2nd edn.
15. Limnios, N., Ouhbi, B. (2006), Nonparametric estimation of some important indicators in reliability for semi-Markov processes, *Stat. Methodol.*, 3, 341–350.
16. Yeh, L. (1997), The rate of occurrence of failures, *J. Appl. Probab.*, 34, 234–247.
17. D'Amico, G. (2015), Rate of occurrence of failures (ROCOF) of higher-order for Markov processes: Analysis, inference and application to financial credit ratings, *Methodol. Comput. Appl. Probab.*, 17, 929–949.
18. Ouhbi, B., Limnios, N. (2002), The rate of occurrence of failures for semi-Markov processes and estimation, *Stat. Probab. Lett.*, 59(3), 245–255.
19. Limnios, N. (2012), Reliability measures of semi-Markov systems with general state space, *Methodol. Comput. Appl. Probab.*, 14(4), 895–917.
20. Votsi, I., Limnios, N., Tsaklidis, G., Papadimitriou, E. (2014), Hidden semi-Markov modeling for the estimation of earthquake occurrence rates, *Commun. Stat. Theory Methods*, 43, 1484–1502.
21. Votsi, I., Limnios, N. (2015), Estimation of the intensity of the hitting time for semi-Markov chains and hidden Markov renewal chains, *J. Nonparametr. Stat.*, 27(2), 149–166.
22. Guédon, Y., Cocozza-Thivent, C. (1990), Explicit state occupancy modelling by hidden semi-Markov models: Application of Derin's scheme, *Comput. Speech Lang.*, 4, 167–192.
23. van der Vaart, A.W. (1998), *Asymptotic Statistics*, Cambridge University Press, New York.
24. Eleftheroglou, N., Loutras, T., Malefaki, S. (2015), Stochastic modeling of fatigue damage in composite materials via non homogeneous hidden semi Markov processes and health monitoring measurements. In: A. Karagrigoriou, T. Oliveira, C. H. Skiadas (Eds), *Statistical, Stochastic and Data Analysis Methods and Applications*, 19–33.
25. Malefaki, S., Limnios, N., Dersin, P. (2014), Reliability of maintained system under a semi-Markov setting, *Reliab. Eng. Syst. Safe*, 131, 282–290.

7

Reliability Measures and Indices for Amusement Park Rides

Stavros Kioutsoukoustas, Alex Karagrigoriou, and Ilia Vonta

CONTENTS

7.1 Introduction 121
    7.1.1 Brief Review 123
7.2 Reliability Theory: Indices and Measures 124
    7.2.1 Basic Reliability Measures 124
    7.2.2 Basic Reliability Indices 126
        7.2.2.1 Mean Time to Failure (MTTF) 126
        7.2.2.2 Availability 126
        7.2.2.3 Maintainability 127
7.3 Applications 128
    7.3.1 Introduction 128
        7.3.1.1 Rotating Electric Train 128
        7.3.1.2 Rotating Dragons 129
        7.3.1.3 Track with Bumper Cars 129
        7.3.1.4 Carousel 130
        7.3.1.5 Drop Zone 131
    7.3.2 Reliability 131
        7.3.2.1 Determination of Time Distributions 133
        7.3.2.2 Variable TM 133
        7.3.2.3 Variable TTF 135
        7.3.2.4 Variable TTR (Four Rides) 136
        7.3.2.5 Variable TTR1 (Track with Bumper Cars) 137
    7.3.3 Availability 138
    7.3.4 Maintainability 139
7.4 Conclusion and Discussion 139
References 140

7.1 Introduction

Amusement parks, also known as *Luna Parks*, are places that every person has either visited or dreams to visit. Amusement parks such as Euro Disney in France, Six Flags in the United States, and Tivoli in Denmark attract thousands of tourists every year (for details, see the International Association of Amusement Parks and Attractions).

Booming technology, together with the growing demand for more and more impressive rides, has created the need for laws, regulations, and standards to ensure the integrity and the construction of an amusement park.

In the context of this work, we make a brief reference to the history of amusement parks, the institutions that led to their creation, the international standards that set the minimum regulations for their safe construction and operation, and fully investigate the reliability measures and indices associated with their operation.

We gathered data from five Luna Park rides from the amusement park Happy Park, based in Kavala, Greece. There were three selection criteria for the rides. Initially, we wanted to choose group rides rather than individual ones. Then, we chose rides that it is possible to find in many Luna Parks today, even if some of them have a history centuries old. The last criterion was the structure of the rides, so that our study does not end up being a one-sided study but has examined as many forms of group rides as possible.

The data collected are real and have been collected from the maintenance books of the rides and refer to the periods during which the machines were maintained and the duration of their maintenance, the damage that each ride had during its operation, and the time spent to restore the damage, as well as the amount of time that it was out of order due to failure/repair.

Next follows a brief description of the reliability tools we are going to use, without expanding on mathematical analyses, and then the actual data are analyzed.

The structure of the rides is the first item to be analyzed, through which the operation and the minimum periods of interruption and operation of the rides are understood. Statistical analysis of data is undertaken to determine the underlying distributions and study the frequency of the failures (D'Agostino and Stephens, 1986; Kalbfleisch and Prentice, 1980).

Finally, the reliability of the rides is calculated on the basis of actual data to understand the relationship between the structure of the ride, its maintainability, and its availability (Frenkel et al., 2014; Lisnianski et al., 2018).

Of course, in all of this, an important role is played by external and unexpected factors such as the weather conditions, the voltage of the electricity system, and the place where the rides operate. All of these external factors cannot be predicted, but it is possible to predict their effect with proper handling, statistical or otherwise, to reduce them to a minimum.

Finally, from the research as a whole, we derive general conclusions about which rides require the greatest attention and which have high availability, and finally provide proposals for improving their performance.

7.1.1 Brief Review

Luna Parks consist of rides that are usually complex constructions with complex operation, aimed at fun and the explosion of adrenaline of the participants. Because of this complexity and that they are made of electromechanical parts, they embody a risk in their installation and operation. There are many accidents involving rides in Luna Parks.

In most European countries, inspection of the rides is legislated for and imposed on the operators of the amusement parks. There are relevant national standards and regulations, as well as harmonized European standards (EN standards) that define the safety requirements that the rides must fulfill concerning design, manufacture, assembly, and operation (e.g., British Standard Institution 1991).

The design, construction, and installation of the rides must be made in accordance with existing European specifications and the mechanisms of the rides must be accompanied by all required certificates. Inspection and the confirmation of compliance with the regulations and standards is the responsibility of accredited bodies for inspection and certification and the licensing competence of state authorities.

The continued development of amusement parks and the accidents that have taken place since the beginning of their operation have created the need for the establishment of safety rules during both construction and operation. In 2004, after 16 years of preparation and with the cooperation of 28 member states of the European Union, the European Standardization Committee (CEN) adopted standard EN 13814, which replaced the German standard DIN 4112 *Fliegende Bauten*. Each country had the right to create its own standard based on EN 13814 and also to apply some national deviations from it.

In this work, methods and techniques about reliability will be presented and studied for five group entertainment rides in Happy Park (http://www.happyparkkavala.gr), which is located in Kavala, Greece. These rides are:

1. Rotating electric train. This is a 16-seat train running on rails of a total length of 75 m, using a 24-volt electric motor, and is aimed at children aged three to eight years old.

2. Rotating Dragons. This consists of six two-seat dragon dummies that rotate around a shaft, and each independently elevates or descends through a hydraulic system by pressing the corresponding button that is placed inside each dragon dummy. It is aimed at children from three to eight years old.

3. Track with bumper cars. This consists of a 10×20 m stage where there are 20 two-seat bumper cars. Movement is possible due to electric motors working from 60 V to 100 V depending on the operator's

setting. It is aimed at children over four years old with a guardian present and over nine years old without a guardian.
4. Carousel. This consists of a circular deck on which there are dummies of horses, elephants, wagons, and so forth, with a total of 32 seats, and is decorated with Renaissance designs and colors. It is aimed at children over three years old with a guardian present and over eight years old without a guardian.
5. Drop Zone. This consists of a tower 14 m high, on which there are 12 seats, three on each side, that go up and down by different rates to cause an adrenaline explosion. It is aimed at children over eight years old.

All the available data (the number of incidents and the repair and preventive maintenance times) used for the present study are real and came from Happy Park, which is located in Kavala, Greece. The data have been collected from the maintenance books that each ride has and records every action that has taken place, either as a result of an emergency repair or due to planned maintenance. Based on these data, we investigate the importance of the reliability of such systems. Among the issues of concern are the determination of the ideal underlying time distribution and the evaluation of the availability and maintainability of the system (Bakouros 2002; Renyan 2015).

7.2 Reliability Theory: Indices and Measures

7.2.1 Basic Reliability Measures

Reliability theory deals with the ability of an object (e.g., a component or unit of a system, machine, or tool, or even an individual) to perform consistently and maintain the required function or task within a specified time period under specific conditions (Hopkins 2000; Verma et al. 2016; Vonta et al. 2008).

In most applications, the possible states of the units of a system at any time point t are usually not known or available. In most cases, the only available information is the probabilities of their operation. In all such cases, we assume that the states of each unit of the system under study are random variables. For a simple two-state system, the states take two possible values, usually denoted by 1 and 0 and representing functioning and failure (O'Connor and Kleyner 2012; Lisnianski and Frenkel 2015). The reliability, denoted by R, is the concept that quantifies the degree and the extent of the operation of a system. Formally, if T denotes the time up to failure, then reliability at the time point t is the probability that T is greater than or equal to t:

$$R(t) = P(T \geq t) = F(t)$$

where $F(t)$ denotes the probability distribution of the random variable T.

The failure or hazard rate, denoted by $h(t)$, represents the instantaneous failure probability within the infinitesimal time interval $[t, t+dt]$ provided that the unit or item is functioning up to time point t. Reliability and failure rate are connected through the expression

$$h(t) = \frac{F'(t)}{R(t)}$$

or equivalently

$$h(t) = \lim_{dt \to 0} \frac{P[t \leq T \leq t+dt \mid T \geq t]}{dt}.$$

where $F'(t) = f(t)$ is the probability density function of T.

The typical form of the failure rate function is called the *bathtub curve*, which appears quite often during the study of a unit's lifetime. In fact, it can be considered to include the case where the variable T expresses the lifetime of a human being or, more generally, of a living being. The most popular distributions used for such type of analysis are the exponential, the Weibull, the lognormal, and the loglogistic.

Basic reliability indices that provide a clear indication of the status of system are:

- Percentage rate of failure/frequency (the rate at which failure occurs)
- Percentage of the total number of product failures during their lifetime
- Median number of failures/damage within predetermined time period
- Median time before failure, denoted by MTBF
- Median time till failure, denoted by MTTF
- Median time till repair, denoted by MTTR
- Availability
- Maintainability

From this, it emerges that MTTR+MTTF=MTBF, while for MTTF we have to specify that it is the expected time until the first failure of the system occurs. If (t) is the time until the failure of the system occurs, then MTTF = $E[T]$ and can be calculated as:

$$\text{MTTF} = E[T] = \int_0^\infty R(t)dt = \int_0^\infty tf(t)dt$$

7.2.2 Basic Reliability Indices

The most popular reliability indices that are presented in this subsection are the mean time till first failure (MTTF) and the availability and maintainability of the system [19].

7.2.2.1 Mean Time to Failure (MTTF)

Recall that T is the random variable representing the time to failure and F is the distribution function of T. Then, the mean time till failure is defined to be the expectation of the random variable T. The relation that connects MTTF with the distribution F, the density f, and the reliability R is given here:

$$\text{MTTF} = E(T) = \int_0^\infty t f(t) dt$$

$$= -\int_0^\infty t d\left[1 - F(t)\right] = -\int_0^\infty t dR(t).$$

By integration by parts, this last equality takes the form

$$\text{MTTF} = -tR(t)\Big|_0^\infty + \int_0^\infty R(t) dt.$$

Finally, this expression reduces to

$$\text{MTTF} = \int_0^\infty R(t) dt$$

in the case where the convergence to zero is faster for the reliability function.

Note that the summation of the MMTFs and MTTRs coincides with the MTBF, since the time between two failures consists of the time till failure plus the repair time required after the first of the two failures to return to the operating state.

7.2.2.2 Availability

Availability is defined as the probability of the system being capable of functioning or conducting its operation when it is needed to do so.

Systems availability criteria refer to systems that can be repaired and short interruptions in their operation are tolerable. Depending on the time horizon that interests us, the availability criteria of the systems are expressed in four different ways.

- The *instantaneous availability* or *point availability* A(t) of a system at time t is the probability that the system will be operational (correctly) at any given time point t.
- The *availability of interval* or *operational availability* A of a system in (0, t) is the expected proportion of time within the operational interval (0, t) in which the system was functioning correctly:

$$A = \frac{\text{System Operating Time}}{\text{System Operating Time} + \text{System Out of Operation Time}}.$$

As expected, the system operating time is related to the reliability of the system while the time during which the system is not functioning is related to the maintainability of the system. The connection between availability, reliability, and maintainability will be presented later.

- The *limiting interval availability* is the expected fraction of the time in which the system operates:

$$\frac{1}{T}\int_0^t A(y)dy$$

- The *steady-state availability* represents the probability of a long period of time during which the system is available. Alternatively, it can be defined as the limit of instantaneous availability:

$$\lim_{t \to \infty} A(t).$$

7.2.2.3 Maintainability

According to ISO/IEC 2382-14, maintenance is a set of activities intended to keep a functional unit in, or to restore it to, a state in which it can perform a required function. Maintenance includes activities such as monitoring, *tests*, measurements, replacements, adjustments, repairs, and in some cases administrative actions.

Maintainability is the ability of a functional unit, under given conditions of use, to be retained in, or restored to, a state in which it can perform a required function when maintenance is performed under given conditions and using stated procedures and resources. (ISO/IEC 2382-14 1997)

A correct and complete maintenance program preserves the equipment at the required level of reliability and operation, increases the availability of equipment when maintenance duration is as short as possible, reduces maintenance costs, reduces equipment wear, and extends its service life. Accurate application of a maintenance program is very important to ensure the optimal operation of the equipment. The types of maintenance are: preventive,

TABLE 7.1

Relationship between the Main Concepts of Reliability Theory

Reliability	Maintainability	Availability
Constant	Decreases	Decreases
Constant	Increases	Increases
Decreases	Constant	Decreases
Increases	Constant	Increases

correctional, predictive, reliability centralized and routine maintenance (Mobley 2004; Rausand and Hoyland 2004).

Table 7.1 provides the relationship between the three main concepts of reliability theory. From the table, we easily deduce that:

- A constant reliability does not necessarily imply high availability.
- A constant reliability (even a low one) could imply high availability provided the maintainability increases (i.e., the time for maintenance decreases).
- If maintainability decreases (i.e., the time for maintenance increases), the availability also decreases.

7.3 Applications

7.3.1 Introduction

7.3.1.1 Rotating Electric Train

The data that we collected for the rotating electric train that are about emergency repair and scheduled maintenance are displayed in Table 7.2. The data

TABLE 7.2

Data for the Ride "Train"

Year	Number of Failures	Time to Repair (Wages' Worth)	Operation Time (Date)	Time Out of Operation
2012		Preventive maintenance of 6 wages' worth		
	0	0	75	0
2013		Preventive maintenance of 10 wages' worth		
	1	1	62	2
		Preventive maintenance 4 wages' worth		
2014	0	0	60	0
		Preventive maintenance of 4 wages' worth		
	2	2	30	1
2015	2	2–4	30	2
		Preventive maintenance of 4 wages' worth		
2016	0	0	60	0

are presented in the order in which they occurred. We can see, for example, that in 2013, the year started with a preventive maintenance period, during which 10 wages' worth of money were spent. It continued with a 62-day period when there was a breakdown that required one wage's worth of money for repair, and the year ended with another preventive maintenance period with four wages' worth of money spent.

7.3.1.2 Rotating Dragons

The Rotating Dragons ride is an adaptation for young children of the ride "Hurricane," built for the first time in 1940 by Allan Herschell Company.

From the maintenance books, we collected the data in Table 7.3. The ride was first put into operation in 2014 and until October 2016 it had only one breakdown, and two preventive maintenance periods (in November 2014 and March 2016).

7.3.1.3 Track with Bumper Cars

From the maintenance and repair books, we collected data for the period 2011–2016, displayed in Table 7.4 in the first four columns. The operation period of the track is about four months per year: three months in the summer and one month during Christmas. We should emphasize that due to the structure of the tracks, during the period for which we have measurements, there has never been a case where it was totally out of operation, since it has a backup system. At the same time, though, the system has occasionally experienced many failures in individual systems. In the fifth column, we have calculated the mean of the failures per year.

We notice that in the first few years, we have about one failure every five days (one failure every 4.5 days in 2011 and one failure every 5.25 days in 2012); then, the number of failures rises rapidly and, at Christmas 2014, we have about one failure every day. The sharp reduction in damage that is seen in 2016 is justified by the radical maintenance and reconstruction that took place on almost all track systems in March 2016.

TABLE 7.3

Data for the Ride "Rotating Dragons"

Years	Number of Failures	Time to Repair (Wages' Worth)	Operation Time (Date)	Out of Operation Time
2014	0	0	60	0
	Preventive maintenance of 6 wages' worth			
	0	0	30	0
2015	1	1	85	1
	0	0	30	0
2016	Preventive maintenance of 8 wages' worth			
	0	0	75	0

TABLE 7.4

Data for the Ride "Track with Bumper Cars"

Year	Number of Failure	Time to Repair (Wages' Worth)	Operation Time (Date)	Rate of Failure. 1 Failure/ Date	Out of Operation Time
2011	Preventive maintenance of 80 wages' worth				
	20	20	90	4.5000	0
2012	Preventive maintenance of 60 wages' worth				
	20	20	105	5.2500	0
2013	34	34	90	2.6471	0
	10	10	30	3.0000	0
2014	42	42	90	2.1429	0
	Preventive maintenance of 60 wages' worth				
	28	28	30	1.0714	0
2015	56	56	90	1.6071	0
	Preventive maintenance 90 wages' worth				
	17	17	30	1.7647	0
2016	10	10	105	10.5000	0

It should be noted that all the failures involved damage that did not prevent the operation of the system (failure to vehicles, lighting, or even the rectifier) and that were often due to unpredictable factors, such as the high humidity in December 2014 that caused mud on the track. As a result, many vehicles had problems with the traction of the brass wheel with the ground.

7.3.1.4 Carousel

In Table 7.5, the collected data from the carousel maintenance books are displayed. The brand-new carousel was put into operation in 2008 and, until 2012, was operating for approximately three months per year without any maintenance or malfunction.

TABLE 7.5

Data for the Ride "Carousel"

Year	Number of Failure	Time to Repair (Wages' Worth)	Operation Time (Date)	Out of Operation Time
2013	Preventive maintenance of 20 wages' worth			
	0	0	90	0
2014	1	6	90	1
	Preventive maintenance of 10 wages' worth			
2015	0	0	90	0
	Preventive maintenance of 30 wages' worth			
2016	0	0	75	0

7.3.1.5 Drop Zone

Table 7.6 contains the data that we have collected from the Drop Zone maintenance books.

7.3.2 Reliability

The Pareto diagram for the total number of wages' worth spent for preventive maintenance for the rides under consideration is presented in Figure 7.1.

TABLE 7.6

Data for the Ride "Drop Zone"

Year	Number of Failure	Time to Repair (Wages' Worth)	Operation Time (Days)
2012	1	2	45
	Preventive maintenance of 20 wages' worth		
	0	0	30
2013	Preventive maintenance of 8 wages' worth		
	0	0	120
	Preventive maintenance of 14 wages' worth		
	0	0	30
	1	2	60
2014	Preventive maintenance of 10 wages' worth		
	0	0	30
2015	Preventive maintenance of 30 wages' worth		
	2	2	75
2016	Preventive maintenance of 8 wages' worth		
	0	0	75

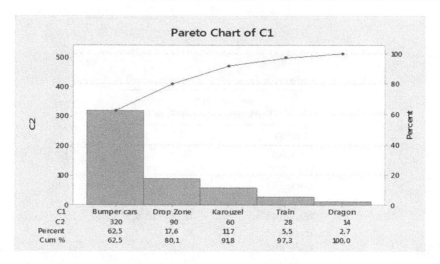

FIGURE 7.1
Pareto diagram of total time preventive maintenance.

From the diagram, the 80–20 Pareto theory is partially confirmed, since 80% of the maintenance costs are due to 40% of the rides (bumper cars and Drop Zone), while 20% of the total maintenance costs are spent on 60% of the rides (carousel, rotating electric train, Rotating Dragons).

In Tables 7.7 and 7.8, the data are grouped according to time out of operation.

In Table 7.8, time till failure (TTF) has not been calculated, since the track with bumper cars had numerous (almost daily) failures until 2015 that make the calculation of TTF meaningless. The situation improved significantly from the middle of 2015, when a 90-wage preventive maintenance spend reduced the failure rate considerably.

TABLE 7.7

Rides with Time Out of Operation > 0

Ride	Group 1 Drop Zone-Carousel-Train-Dragon		
	TTF	TTR	FAIL
Drop Zone	22	1	1
Drop Zone	30	2	1
Drop Zone	25	1	1
Drop Zone	25	1	1
Carousel	30	6	1
Train	31	1	1
Train	10	1	1
Train	10	1	1
Train	10	4	1
Train	10	2	1
Rotating Dragons	42	1	1

TABLE 7.8

Ride with Time Out of Operation = 0

CLASS1	Group 2 Track with Bumper Cars	
	TTR	FAIL
Track 2011	20	20
Track 2012	20	20
Track 2013	34	34
Track 2013	10	10
Track 2014	42	42
Track 2014	28	28
Track 2015	56	56
Track 2015	17	17
Track 2016	10	10

The Pareto charts associated with Tables 7.7 and 7.8 appear in Figure 7.2. The charts clearly show the most efficient rides, the ride with the most failures, and the behavior over time of the track with bumper cars.

The charts in Figure 7.2 fail to follow the 80%–20% Pareto theory. Instead, a nearly 70%–30% analogy appears to apply, which could be attributed to the presence of a number of special characteristics. We should point out something that cannot be easily understood by the observation of the tables. For instance, the most costly ride, in repair wages' worth spent, is the rotating train, while the Rotating Dragons are the most economical, with only one wage's worth spent for the period under investigation. The causes and the effects of these issues may be hidden in flow charts.

Furthermore, the flow chart of the Rotating Dragons is mostly parallel, and this allows for smaller sets of operation groups and fewer failures, which make it more flexible in the event of a failure of a building block. The carousel has also a parallel operating system and it can operate even when some of its building blocks are not functioning. In contrast, the train has a serial flow chart that does not allow failure of a building block. The Drop Zone has also a serial operating structure with many critical components that make it vulnerable to damage. In addition to the operating structure, age, complexity, weather conditions, and preventive maintenance are key issues for causing or avoiding failures.

The Pareto chart for the track with bumper cars also does not comply with the 80%–20% theory. We observe a stable number of failures (2011 and 2012) followed by a constant increase in failures until 2015. The complete preventive maintenance and reconstruction that took place at the end of 2015 resulted in a great reduction of failures in 2016. Another factor that significantly affects the number of failures per year is the total number of days that the track was in operation. We observe that before 2013, the track operated at most for 105 days per year, while since 2014 it operates for 120 days per year.

7.3.2.1 Determination of Time Distributions

After grouping the data and plotting them with the Pareto chart, we will continue our study trying to find the right distribution for the variable time of maintenance (TM) for all five rides, for the variable TTF, for the variable time till repair (TTR) for four of the rides, and for the variable TTR for the track with bumper cars.

7.3.2.2 Variable TM

By observing Figure 7.3, the lognormal distribution is the one that best fits the TM data, with the loglogistic coming very close (Anderson-Darling test statistic = 1132 for lognormal and 1162 for loglogistic). The good fit of the lognormal distribution is also evident from Figure 7.4, where the parameters of the distribution have been estimated (location = 2.70443, scale = 0.977522).

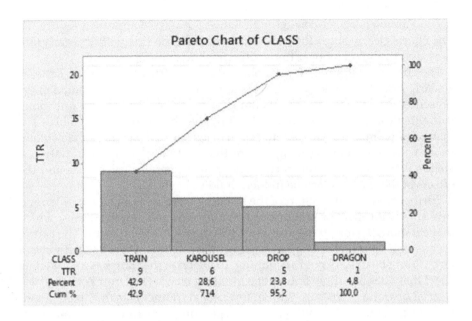

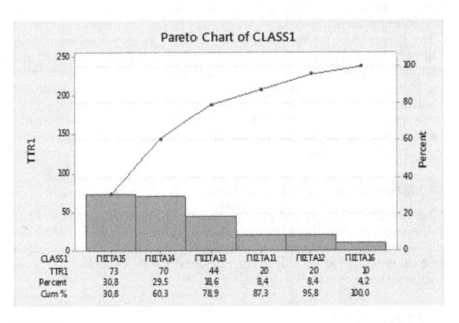

FIGURE 7.2
Pareto chart of TTR for the rides.

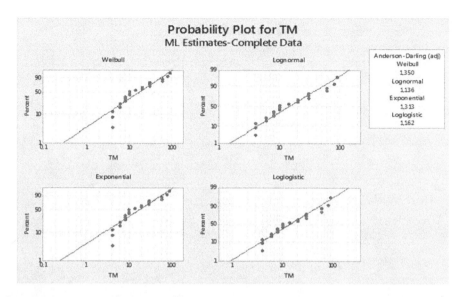

FIGURE 7.3
Customization of TM data into distributions.

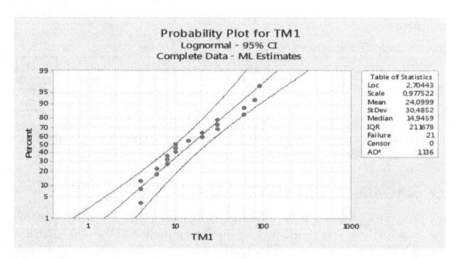

FIGURE 7.4
Parametric lognormal probability analysis of variable TM.

7.3.2.3 Variable TTF

To study the distribution of the TTF variable, we follow the same procedure as before. The results are presented in Figure 7.5. We observe that the distribution of TTF cannot be perfectly identified. However, the distribution that seems to be closer to the data is the Weibull (AD test statistic = 1.761), followed closely by the loglogistic distribution (AD test statistic = 1.819). For

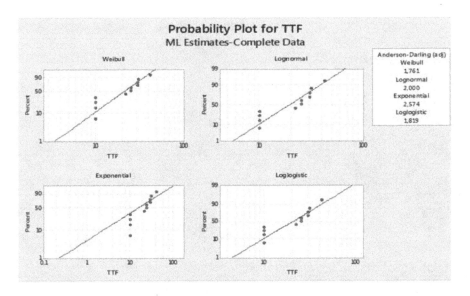

FIGURE 7.5
Customization of TTF data into distributions.

TABLE 7.9

Distributional Characteristics of TTF

	Shape/Location	Scale	Mean	Std	AD Test
Weibull	2.316	25.238	22.36	10.247	1.761
Loglogistic	3.011	0.329	24.43	19.299	1.819

comparative purposes, Table 7.9 displays the basic parameter characteristics of both distributions.

7.3.2.4 Variable TTR (Four Rides)

For four rides (rotating train, Rotating Dragons, carousel, and Drop Zone), the same analysis was conducted for the determination of the ideal distribution of TTR. The results of the analysis are less promising than before, since none of the well-known distributions provides a fit that can be considered very satisfactory. For the sake of completeness, the distributional characteristics of the best two fits (Weibull and exponential) are presented in Table 7.10.

TABLE 7.10

Distributional Characteristics of TTR (Four Rides)

	Shape	Scale	Mean	Std	AD Test
Weibull	1.409	2.126	1.936	1.393	2.404
Exponential	1.000	1.910	1.910	1.910	2.475

7.3.2.5 Variable TTR1 (Track with Bumper Cars)

Following the same procedure as before, we identify the best distribution for the time to repair (TTR1) for the track with bumper cars. Figure 7.6 shows that the best fit is attained with the loglogistic distribution (AD test statistic = 1.529) followed closely by Weibull (AD test statistic = 1.556).

The good fit of the loglogistic distribution is also evident from Figure 7.7, where the distributional characteristics of the distribution are also displayed (location parameter = 3.120, scale parameter = 0.338).

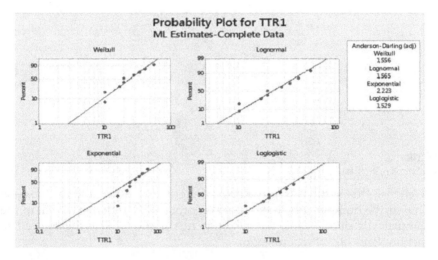

FIGURE 7.6
Customization of TTR1 data into distributions.

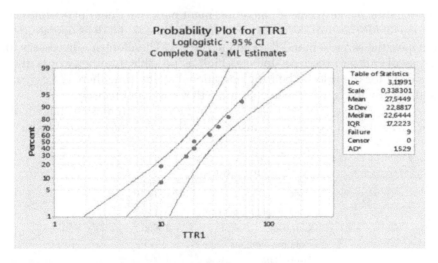

FIGURE 7.7
Parametric analysis of the loglogistic distribution of TTR1.

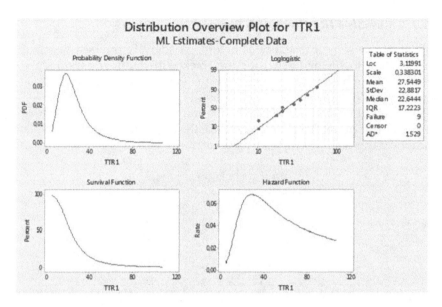

FIGURE 7.8
Functions of TTR for bumper cars.

We close this section by providing, in Figure 7.8, especially for the time to repair the bumper cars, the associated probability function, the fit of the loglogistic distribution, and the reliability function, as well as the failure rate function (Peto 1973; Turnbull 1976).

7.3.3 Availability

Availability is a key issue for an amusement park that does not operate continuously for the entire year and, even for the time that it does operate, does not have the same demand throughout the day. Therefore, it is necessary for the availability of the rides to be as high as possible. In other words, there is a need for the rides to be fully operational at any time there is a demand without anyone being able to determine that moment precisely.

In Table 7.11 we observe that, although the ride with most damage is the track with the bumper cars, it is also the one that was always available during

TABLE 7.11

Availability of Rides

Ride	Availability
Track with bumper cars	1
Carousel	0.9956
Rotating Dragons	0.995
Drop Zone	0.992
Rotating train	0.97

the period under study, with a perfect rate of 100% availability. The carousel, with 99.56%, and the Rotating Dragons, with 99,5%, have the next highest availabilities. Finally, the Drop Zone and the rotating train have availabilities of 99.2% and 97%, respectively.

7.3.4 Maintainability

Maintenance is a set of actions designed to retain the functionality of a unit or item and restore damaged components to a level so that it again becomes capable of executing its function. Maintainability is the possibility, for a given operating condition, to perform a maintenance process or a successful repair within a given time period. In other words, maintainability is the speed at which a system returns from malfunctioning to full operation.

Luna Parks rides consist of electromechanical equipment that must be able to operate in specific time periods, which, at times, may be hours or even days. For this reason, maintainability should be as limited as possible or, if possible, non-existent. As maintainability decreases, the role and significance of preventive maintenance becomes more evident.

During preventive maintenance, everything that may cause a problem or malfunction during operation may be prevented and the equipment maintained. The maintenance of the amusement park's rides should be based on the specifications of the rides, the experience of their operators, the instructions of the inspectors, and the operating time of each ride.

The target of maintenance is for the equipment's reliability and availability to be at the highest levels, to keep it safe for both the occupants and the operators, restore the proper functioning of the equipment, and, at the same time, eliminate causes of failure and keep the operation and maintenance costs as low as possible. The types of maintenance can be categorized into the following:

- Preventive maintenance
- Correctional maintenance
- Predictive maintenance
- Reliability centered maintenance
- Routine maintenance

7.4 Conclusion and Discussion

In this chapter, we thoroughly discussed statistical quality control and reliability, maintainability, and availability of amusement parks. Some brief concluding remarks are stated here:

TABLE 7.12

Ranking of Rides Based on Cost for Preventive Maintenance

Ride	Wages' Worth for Preventive Maintenance	Percentage (%)
Track with bumper cars	320	62.5
Drop Zone	90	17.6
Carousel	60	11.7
Rotating train	28	5.5
Rotating Dragons	14	2.3

We observe that for the period that we studied, the aggregated maintenance costs are shown in Table 7.12.

We observe that the ride with the highest maintenance requirements is, by far, the track with bumper cars, with a long gap to the second one, the Drop Zone. This, apart from the ride's age, may also be due to its complexity, since each of the 20 cars that it contains has the same maintenance needs as a simple ride, since it contains a motor, a voltage extraction system, a wheel, a steering system, and so on. It is also the ride that is most affected by the weather conditions. Rainy weather brings moisture and mud onto the track via the visitors' shoes and creates problems with the track connections. The Drop Zone comes second in maintenance requirements, and we could say that its needs are defined by its complexity, not only its serial structure but also by the many security measures it has to follow. Third in terms of maintenance cost is the carousel. The other two rides, the rotating train and the Rotating Dragons, are the most economical.

Maintainability may be improved if one knows the components that appear to experience frequent failures and stores a few of the most common spare parts needed for their operation. Finally, one must be reminded that the operation of any machinery, not only that which involves children, should be operated according to the manufacturer's manual and maintenance should be performed according to the manufacturer's specifications. Such actions will ensure an improved maintainability and, consequently, a satisfactory reliability and availability.

References

Bakouros I (2002) *Reliability and Maintenance* (in Greek), Hellenic Open University, Patras, Greece.

British Standards Institution (1991) *BS4778-3.1:1991 Quality Vocabulary. Availability, Reliability and Maintainability Terms. Guide to Concepts and Related Definitions*, British Standards Institution, London.

D'Agostino RB and Stephens MA (1986) *Goodness-of-Fit Techniques*, Marcel Dekker, New York, NY.

Frenkel I, Karagrigoriou A, Lisnianski A, and Kleyner A (2014) *Applied Reliability Engineering and Risk Analysis: Probabilistic Models and Statistical Inference*, Wiley Series in Quality and Reliability Engineering, John Wiley and Sons, New York.

Hopkins WG (2000) Measures of reliability in sports medicine and science, *Sports Medicine* 30 (1): 1–15.

International Association of Amusement Parks and Attractions, https://www.iaapa.org.

ISO/IEC 2382-14 (1997) Information technology: vocabulary: Part 14: Reliability, maintainability and availability, https://www.iso.org/obp/ui/#iso:std:iso-iec:2382:-14:ed-2:v1:en.

Kalbfleisch JD and Prentice RL (1980) *The Statistical Analysis of Failure Time Data*, John Wiley & Sons, New York, NY.

Lisnianski A and Frenkel I (2015) *Recent Advances in System Reliability: Signatures, Multi-State Systems and Statistical Inference*, Springer, Berlin.

Lisnianski A, Frenkel I, and Karagrigoriou A (2018). *Recent Advances in Multi-State Reliability, Theory and Applications*, Springer Series in Reliability Engineering, Springer, Berlin.

Mobley KR (2004) *Maintenance Fundamentals*, 2nd edn, Elsevier.

O'Connor PDT and Kleyner A (2012) *Practical Reliability Engineering*, 5th edn, Wiley, Chichester, UK.

Peto R (1973) Experimental survival curves for interval-censored data, *Applied Statistics* 22: 86–91.

Turnbull BW (1976) The empirical distribution function with arbitrarily grouped, censored and truncated data, *Journal of the Royal Statistical Society* 38: 290–295.

Rausand M and Hoyland A (2004) *System Reliability Theory: Models, Statistical Methods, and Applications*, 2nd edn, Wiley, Hoboken, NJ.

Renyan J (2015) *Introduction to Quality and Reliability Engineering*, Springer, Science Press, Beijing.

Verma AK, Srividya A, and Karanki DR (2016) *Reliability and Safety Engineering*, 2nd edn, Hoang Pham (ed.), Springer, Piscataway, NJ.

Vonta F, Nikulin M, Limnios N, and Huber-Carol C (2008) *Statistical Models and Methods for Biomedical and Technical Systems*, Springer (Birkhäuser), Berlin.

Westgard JO, Groth T, Aronsson, T, Falk H, and de Verdier CH (1977) Performance characteristics of rules for internal quality control: Probabilities for false rejection and error detection, *Clinical Chemistry* 23 (10): 1857–1867.

8

On Parameter Dependence and Related Topics: The Impact of Jerzy Filus from Genesis to Recent Developments

Jerzy Filus, Lidia Filus, Barry C. Arnold, Pavlina K. Jordanova,
Ludy Núñez Soza, Ying Lu, Silvia Stehlíková, and Milan Stehlík

CONTENTS

8.1 Introduction: A Literature Review 144
    8.1.1 The Beginnings 144
    8.1.2 Rediscovery and Generalization of Freund (1961) 145
    8.1.3 Load Optimization 145
    8.1.4 Work at Illinois Institute of Technology and the Beginnings of an Alternative to Copulas 146
    8.1.5 Modification of the Multidimensional Marshall and Olkin Model 149
8.2 Extension of Parameter Dependence Paradigm to the General Case of Stochastic Dependence 149
8.3 Pseudoexponential Models in Medicine 152
    8.3.1 Application of Pseudoexponential Models to Fractal Hypothesis for Cancer 152
    8.3.2 Application of Pseudoexponential Models to Dose Finding Studies 155
    8.3.3 Application of Pseudoexponential Models to Bone Mineral Density 155
8.4 Relationship of Pseudo-Weibull and Marshall–Olkin Weibull Models: Application to Administrative Arsenic Data from Parinacota, Chile 157
8.5 Discussion by Pavlina Jordanova 160
8.6 Discussion by Barry C. Arnold 164
8.7 Reply by Jerzy Filus to Barry C. Arnold's Comments 166
8.8 Rejoinder on Joiners by Barry C. Arnold 168
References 169

8.1 Introduction: A Literature Review

In this section, via an interview with Jerzy Filus discussing his career from his early years in Poland all the way up to today, we will unfold and explore the genesis and development of multivariate pseudonormal distributions, parameter dependence, and finally stochastic dependence in general, which will be fully discussed in Section 8.2. The medical applications of pseudo-exponential models are discussed in Section 8.3, while in Section 8.4 a relationship between the pseudo-Weibull distribution and the Marshall–Olkin model is presented. Discussion on these and other related issues by Pavlina Jordanova and Barry C. Arnold appear in Sections 8.5 and 8.6.

8.1.1 The Beginnings

Milan: Jerzy, it is nice to have you here, at your workshop "Stochastic Dependence and Related Topics" in Valparaíso. Can you remember your early scientific years?

Jerzy, Lidia: In 1973, Jerzy started to work in the research institute of the Polish Academy of Sciences, later named the System Research Institute. Since the start, he devoted himself to applied probability, even though, originally, his areas of interest were set theory and abstract algebra. In 1976, his new boss, Professor Stanislaw Piasecki, started to give recognition to Jerzy's research, associated with multivariate probability distributions. After Jerzy had finished a talk at the departmental seminar in the fall of 1976 describing his research, Piasecki said to him, "Start to write your PhD thesis." For Jerzy, at that time this was an unexpected and shocking statement. In less than two months, however, the first version of the thesis was ready. Piasecki liked it very much. Jerzy's wife Lidia mentioned this story to her PhD supervisor, Professor Los, who was very well known in the world of mathematicians. He asked her to bring him Jerzy's paper to see it, just for curiosity. He also liked it, and said that Jerzy might need some more support, but he could not help a lot because he was from a different area of mathematics. However, to help "Lidia's husband," he offered to send a copy of the paper to his friend from Wroclaw, professor Ryll-Nardzewski, who at that time was a very prominent mathematician, very well known, not only in Poland but worldwide. Besides, unlike Professor Los, he was a specialist in probability theory. Ryll-Nardzewski also liked Jerzy's PhD paper very much and made an appointment with him at Warsaw University (in the Palace of Culture and Science). They discussed for more than 45 minutes. The main subject was the second part of Jerzy's thesis having a combinatorial character associated with a general reliability structure of an arbitrary system. Ryll-Nardzewski was talking mainly about his own ideas that came to his mind after reading the paper. Meanwhile, a small crowd of people formed at the door. Everyone wanted to talk for a few minutes with the "famous guy." Lidia, who was waiting for Jerzy, repeated later what they were talking about. At one moment a man loudly complained, "What is this 'big man' talking to him about for so long??" These events and the names "Los" and "Ryll-Nardzewski" changed Jerzy's position as a researcher dramatically. He now became a "prominent young scientist," known as "very good" not only in the institute but also among scientists from other institutes and schools all over Poland. The future looked promising.

8.1.2 Rediscovery and Generalization of Freund (1961)

Milan: How does the work of Jerzy relate to the results of Freund (1961)?

Jerzy, Lidia: In 1979/80, Jerzy and Lidia were on a postdoctoral research fellowship at CORE in Louvain-la-Neuve, Belgium, finally having access to a good scientific library. At that time in Poland, there were problems in finding Western-published articles; therefore, it is not surprising that Jerzy rediscovered the results of Freund (1961) in his PhD thesis (Filus 1978) written in 1976–1978 (System Research Institute of the Polish Academy of Sciences, 1979). The story follows that in summer 1980, Jerzy participated in a statistical conference in Cambridge, England, and after his talk he was privately asked a question by one person from the audience, "Have you read Freund?" He was surprised, because he never had heard about Freund and confused him with Austrian psychoanalyst Sigmund Freud; therefore, he replied, "Not very much." Then the man from the audience asked again, "What have you read by Freund?" and Jerzy replied: "Oh, just *Lectures from Psychoanalysis* (Freud), fragments of books of some of his followers, a book by Clara Thompson (*Psychoanalysis: Evolution and Development*) plus some others, but I do not remember a lot. It was a long time ago." At that moment, he was in consternation when he saw the reaction of the man.

However, Jerzy not only rediscovered, but also generalized the results of Freund. Freund (1961) formulated his results for exponential distributions only, while Jerzy, in his PhD thesis (1978), considered the same physical mechanism for all distributions with hazard (failure) rates. Moreover, Freund (1961) was working with bivariate distributions only and Jerzy (1978) considered the general multivariate case. This thesis (in Polish) was never published.

8.1.3 Load Optimization

Milan: Jerzy, you mentioned once that you had contacted reliability groups in the United States. Can you give us more details?

Jerzy, Lidia: In 1980, during his visit to Louvain-la-Neuve, Jerzy found papers by Taylor (1979), Phoenix, and some other authors from Cornell University working on the grant associated with determining the reliability of transoceanic bundles of fibers (based on the reliabilities of single fibers) for telecommunications. He found this (when a number of fibers were considered small) similar to his PhD results, and thus contacted Taylor, Phoenix, Barlow,

Proschan, and probably also Birnbaum. All of them answered positively. In summer 1980, both Barlow and Proschan invited Jerzy to their departments. Barlow wrote, "Come any time." Also, Taylor and Phoenix invited Jerzy to visit Cornell University. Jerzy visited the United States in March 1981 and for some reason this was too late. However, he visited Cornell and met Taylor. As it turned out, at that time Phoenix was having a sabbatical at California Institute, close to San Francisco. Thus, Jerzy visited San Francisco and, in his hotel room, met Phoenix who arrived there from his institute. They had two hours of very fruitful discussion, which was a turning point in Jerzy's research. In particular, the key point for Jerzy was Phoenix's opinion that "a parameter of exponential probability distribution of an object's (like a fiber) lifetime should be considered a function of the stress." He also advised application of power functions with high exponents (even in the range of about 80). As the first result of Phoenix's suggestion, Jerzy wrote his recently frequently cited papers on load ("stress") optimization of repairable systems (Filus 1986, 1987). Jerzy presented this (already published) idea in spring 1989 at a reliability conference in Washington, D.C.; however, there was no feedback, which was very discouraging, and Jerzy dropped this subject. About the same time (in 1988) he discovered the *method of parameter dependence* for the construction of multivariate probability densities. This method was related to the idea of the load optimization, but was different. Also, regardless of the similarities, it was different from the *conditioning method* for the multivariate distribution construction as developed by Arnold, Castillo, Sarabia et al. (1993). The classes of the models obtained by the conditioning method and by the method of parameter dependence are essentially disjoint. As a kind of fusion of the two approaches, the common paper Filus, Filus, and Arnold (2010) demonstrated a method of extension of the bivariate normal density where only six parameters were needed; in the Arnold et al. approach, eight parameters were needed as a minimum.

The method of parameter dependence turned out to be very fruitful and resulted in countless numbers of new models and many works produced until recently. It also yields to the construction of many stochastic processes with both discrete and continuous time. All that had its roots "in the hotel room in San Francisco."

8.1.4 Work at Illinois Institute of Technology and the Beginnings of an Alternative to Copulas

Milan: Jerzy, you worked at IIT [Illinois Institute of Technology] in Chicago, where you met Abe Sklar and Maurice Frank. Recently, copulas have become very popular, and there have been discussions and papers on copulas generally (Mikosch 2005), in support of copulas (Stehlík 2016 and Sumetkijakan 2017 and references therein). A typical copula (in the sense of Baire categories) is purely singular (Trutschnig and Fernandez-Sanchez 2014). At the same time, you developed your models. Do you remember that time?

Lidia, Jerzy: Jerzy developed these pioneering ideas while working at Illinois Institute of Technology, Chicago, where he was recognized several times by his department chair Prof. Maurice J. Frank who said to him, "How are you so lucky to get such nice results?" or, publicly, "Filus may have good results." At that time, Sklar was also a member of IIT and he referred Jerzy to his paper on copulas. At IIT Mathematics Department seminars, Jerzy gave talks on his first results involving parameter dependence. The first talk was given in fall 1988. At that time, bivariate and multivariate pseudonormal distributions had already been constructed but the alternative method of construction by pseudotranslation transformation was only discovered later in December 1988. One month later, pseudotranslations were extended to pseudoaffine transformations as the tool for construction. This was then presented in January 1989 at the Mathematics Department of UIC, Chicago. Parts of his (already developed) work were communicated by Jerzy to Samuel Kotz in 1993 and 1994. Kotz's answers were positive and he especially liked the theorem according to which the class of all the multivariate pseudonormal distributions was invariant with respect to all pseudoaffine transformations that form a group. Then, after a long break, in 1997 Samuel Kotz asked Jerzy to write a short version of what was most important in his results, as he was preparing a new edition of the book *Continuous Multivariate Distributions* (Johnson and Kotz 1972). Kotz also recommended that he submit a manuscript to the *Pakistan Journal of Statistics (PJS)* where the paper appeared in 2000 (Filus and Filus 2000). Also, Jerzy and Lidia were invited to a statistical conference in Lahore, Pakistan, in 2001, where their results were acknowledged, and they decided to publish a series of papers on "pseudodistributions". Subsequently, however, in the years 2001–2003, they had to extend the idea that the pseudonormal construction was based on to all other distributions having probability densities that depend on a parameter (Filus and Filus 2001). This, unexpectedly, turned out not to be an easy task but was finalized successfully and they published most of those results in *PJS* in Filus and Filus (2006). Meanwhile, the transformation approach to the construction of the multivariate pseudonormal distributions was cited in Kotz, Balakrishnan, and Johnson (2000), on pp. 217–218. This was, to the best knowledge of the authors, the first application of the construction of the pseudoaffine transformations in their first "primitive" form. This version of the transformation method, as cited in Kotz, Balakrishnan, and Johnson (2000), was introduced as follows: Let $T_1,...,T_k$ be independent normally distributed random variables. The following pseudoaffine transformation is applied to them:

$$X_1 = aT_1$$

$$X_2 = \phi_1(T_1)T_2 + \theta_1(T_1)$$

$$X_3 = \phi_2(T_1,T_2)T_3 + \theta_2(T_1,T_2)$$

$$X_k = \phi_{k-1}(T_1,...,T_{k-1})T_k + \theta_{k-1}(T_1,T_2,...,T_{k-1})$$

Here, $a \neq 0$ is a real number, $\phi_i(.), i = 1,..,k-1$ are continuous real functions that are never zero, while $\theta_i(.)$ are real and continuous only (in particular, they may take on a value of zero or be zero everywhere). For this transformation, the inverse and its Jacobian can readily be obtained. The random vector so obtained $(X_1,...,X_k)^T$ has pseudonormal distribution and any k-variate normal distribution is a special case of the pseudonormal. If $T_1,...,T_k$ have any pseudonormal distribution, then the resulting random vector also has some pseudonormal distribution. The pseudonormal distributions, as obtained here by means of the pseudoaffine transformation, may as well be obtained by the method of parameter dependence. In Filus and Filus (2000, 2001), the authors gave a reliability motivation for pseudonormal distributions using the parameter dependence paradigm. The essence of the reliability interpretation for the pseudonormals (and later for other similarly motivated "pseudodistributions") or the preconceptions of such an approach can be recognized in Jerzy's papers (Filus 1986, 1987). The parameters of the exponential and gamma lifetime distributions considered in those papers were assumed to be functions of the load put on such objects as trucks (cargo). However, the optimization problem considered (to find an optimal value of the cargo) did not include any construction of the bivariate distributions. The load considered was a controllable deterministic quantity but not a random variable. Even earlier, in Filus and Piasecki (1980), the preconception of parameter dependence was present but the considered distributions were more Freund-like, so the difference is significant. While at IIT, contact with Sklar and his idea about copulas inclined Jerzy to think how to improve his pseudodistributions obtained by the method of parameter dependence. Unlike the copula paradigm, they all had one disadvantage. Physically (in the bivariate case, for example), they all describe the situation where only one object affects another, but this second object does not (physically) affect the first. The situation is perfect for constructions of high-dimensional distributions, especially stochastic processes, but when modeling the reliability of multicomponent systems it is hard to believe in the assumption that there is one way that the components impact. After he quit working at IIT, Jerzy spent more than 10 years trying to find the clue to stochastic models that would comprise mutual physical interactions between physical objects (system components, for example). Finally, he succeeded in this around 2005. The bivariate distributions he had obtained have a possibly universal form for the bivariate survival functions; now, probably, this form may be considered as an alternative to the copula approach. Originally, the main pattern of the construction (see Filus and Filus 2017b) relied on conditioning such as that by Aalen (1989), who modified the conditioning invented by Cox (1972). Later, it turned out that even for the wider class of bivariate distributions one can apply the method of construction independent from Aalen's conditioning or any other. According to that (possibly the newest) method, any two univariate survival functions can be connected in a variety of ways by means of the so-called *joiners*, which are types of dependence functions. At the moment, it looks like any bivariate survival function (not necessarily having a corresponding density) can be obtained in that way. Under some general condition for the joiner, the two univariate distributions originally given become marginal in the constructed bivariate. These quite recent results have not yet been published.

8.1.5 Modification of the Multidimensional Marshall and Olkin Model

Milan: Jerzy, what about the relation of your work to the Marshall and Olkin model?

Lidia, Jerzy: A little apart from the mainstream of Jerzy's scientific activity, associated with the method of parameter dependence and its derivatives, was an episode associated with his modification of the Marshall and Olkin (1967a, 1967b) model. In summer 1987, Jerzy presented some of his PhD thesis's (Filus 1978) ideas at a statistical conference in Pittsburgh, Pennsylvania, where both Marshall and Olkin were present for his talk. He had a short conversation with Olkin on his past attempts (in the mid-1970s) to modify the Marshall and Olkin model. Olkin had shown an interest and asked Jerzy to send him a draft of that research. However, it was yet far from ready and Jerzy was able to send Olkin the draft only in the middle of 1988, at which time the correspondence between the two started. Olkin liked Jerzy's ideas and they were finally published in Filus (1990). The paper contains some modifications of the multidimensional Marshall and Olkin model. Previously, he had also sent the manuscript to Samuel Kotz, announcing that he had probably discovered a new probability distribution. Kotz answered, "You did not obtain a new distribution but a bunch of new probability distributions." However, the paper from 1990 was not actually recognized by the statistical society. Jerzy dropped this subject and now concentrated on the parameter dependence idea and its derivatives such as the construction of stochastic processes, modeling the mutual physical dependence, and others.

8.2 Extension of Parameter Dependence Paradigm to the General Case of Stochastic Dependence

Here, we discuss some more general mechanisms of stochastic dependence, comparing them with the parameter dependence-related constructions. First, let us explain how the general pattern of dependence differs from parameter dependence, while at the same time parameter dependence is a part of the general scheme. Thus, in all the cases considered, including the parameter dependence constructions, a presence of some, say, explanatory variables $X_1,\ldots,X_k$ cause a change to the original (baseline) hazard rate $\lambda_0(y,\theta)$ associated with a random variable Y of the main interest, where θ is a scalar or a vector parameter.

To illustrate the general situation, suppose that a random variable Y is a residual lifetime of a person, given some fixed age, who was diagnosed with some kind of cancer and the question is: What is the probability that the patient, given a treatment, will survive for at least the next five years, that is, $P(Y \geq 5)$?

Suppose that, based on previous experience, it is known that if the person was not a smoker nor an excessive drinker, nor was subjected to other stresses such as prison, war, and so forth, his hazard rate $\lambda_0(y,\theta)$ is the baseline and, according to that, the probability is, say, 0.7. Suppose that we have good enough measures of the stresses (with realizations varying from patient to patient) $X_1,\ldots,X_k$. We assume that any set of realizations $x_1,\ldots,x_k$ of the random variables $X_1,\ldots,X_k$ causes a proportional change of the hazard rate $\lambda_0(y,.)$ to a different hazard rate, say, $\lambda(y,.)$. Having the measurements $x_1,\ldots,x_k$ for the given patient and supposing that the patient was a smoker, consuming a given amount of nicotine per day for 20 years, it may turn out that for him/her, the probability of surviving at least the next five years is 0.4. To measure such varying probabilities, we need a proper stochastic model (as well as a proper method of defining and measuring stresses such as $x_1,\ldots,x_k$). In general, the random variable Y can be arbitrary, but in applications it mostly means the lifetime of any biological, technical, or other object.

However, in cancer treatment applications, Y may also be a size (diameter or area) of a tumor that stochastically depends on stresses. Now, let us analytically describe several dependence mechanisms, all subjected to the same general pattern of changing the hazard rate given the stresses. Given realizations $x_1,\ldots,x_k$ of, generally speaking, explanatory variables $X_1,\ldots,X_k$, the change of the baseline (all the stresses are zeros) hazard rate $\lambda_0(y,\theta)$ to another value, say $\lambda(y,\theta)$, may be realized in several different ways. One of the ways the hazard rate changes is that considered by the method of parameter dependence. In this, we consider the change of the parameter θ to its new value, say $\theta^* = \theta^*(x_1,\ldots,x_k)$, which continuously depends on a realization $(x_1,\ldots,x_k)$ of the random vector $(X_1,\ldots,X_k)$ (see Filus and Filus 2013). This procedure results in the conditional hazard rate $\lambda(y\,|\,(x_1,\ldots,x_k)) = \lambda(y;\theta^*(x_1,\ldots,x_k))$ and, in this way, the conditional probability distribution of $Y\,|\,(x_1,\ldots,x_k)$ is defined also.

Multiplying this conditional distribution by the joint probability distribution of the explanatory random variables $X_1,\ldots,X_k$, one obtains the joint probability distribution of the random vector $(Y,X_1,\ldots,X_k)$. In particular, when $k=1$ and $X_1 = X$, the latter describes the method of construction of bivariate distributions of (X,Y) by parameter dependence. Of course, change of the parameter θ is not the only way the baseline hazard rate may change.

Another two possibilities of change to $\lambda_0(y,\theta)$, as caused by occurrences of quantities $x_1,\ldots,x_k$, are those described by the Cox (1972) model and by its modification by Aalen (1989). In all three cases, some way of conditioning of Y, given $x_1,\ldots,x_k$, is defined.

Thus, in the case of the Cox model, the parameters θ of the hazard rate do not change, but instead the baseline $\lambda_0(y,\theta)$ is multiplied by a factor that is a function of the stresses $x_1,\ldots,x_k$. as caused by occurrences of quantities. Aalen (1989) modifies the Cox paradigm and, instead of multiplying the hazard rate by a function of stresses, adds to it a term that is also a function of stresses. The Aalen approach has some advantages over the Cox approach, since his model, when it is applied to the construction of, say, bivariate survival functions, has some property that may be considered a symmetry. This symmetry relies on the possibility of mutual physical dependence; that is, each of the two random variables, say, X, Y, can be considered to be the explanatory variable for the other. Let us explain this more closely. To do so, it is convenient first to consider the following bivariate survival function, which is a part (the special case) of the final model:

$$S(x,y) = P(X \geq x, Y \geq y)$$
$$= \exp\left[-\int_0^x \lambda_1(t)dt - \int_0^x\int_0^y \psi(t;u)dudt - \int_0^y \lambda_2(u)du\right] \quad (8.1)$$

The latter may be considered to be derived from the Aalen conditioning. Realize, however, that this (final) model is the joint survival function, so corresponding to the random events $(X \geq x, Y \geq y)$, while all the models obtained by the method of parameter dependence are joint densities, so they correspond to the elementary events, say, $(X = t, Y = u)$. Thus, the Aalen conditioning corresponding to an event $(X = t, Y \geq y)$ here has the form of the following conditional hazard rate:

$$\lambda_1(t \mid Y \geq y) = \lambda_1(t) + \int_0^y \psi(t;u)du$$

Similarly, we have:

$$\lambda_2(u \mid X \geq x) = \lambda_2(u) + \int_0^x \psi(t;u)dt$$

So, the hazard rates $\lambda_1(t), \lambda_2(u)$ are the original baseline hazard rates of the considered random variables X, Y. The (original) Aalen terms can be represented here by

$$\int_0^y \psi(t;u)\,du = A_x(t,y)$$

$$\int_0^x \psi(t;u)\,dt = A_y(x,u)$$

Notice that, with the exponentiality assumptions $\lambda_1(t) = \lambda_1, \lambda_2(u) = \lambda_2$ and the condition $\psi(t;u) = a =$ constant, one obtains from Equation 8.1 the following familiar bivariate first exponential Gumbel (1960) distribution: $S(x,y) = P(X \geq x, Y \geq y) = \exp[-\lambda_1 x - axy - \lambda_2 y]$. The latter is a very special case of the models considered here. On the other hand, the bivariate distribution Equation 8.1 can easily be generalized for the case when at least one hazard rate does not exist and the function $\psi(t;u)$ does not exist either. This general bivariate takes the form:

$$S(x,y) = P(X \geq x)P(Y \geq y)\exp[-\alpha(x,y)]$$

The latter form of bivariate survival function, given the original (baseline) survival functions $P(X \geq x), P(Y \geq y)$ is, probably, universal and, as such, may be an alternative for the representation of copulas. More on this subject, together with the conditions the triples $\lambda_1(t), \lambda_2(u), \psi(t;u)$ (when they exist) must satisfy for $S(x,y)$ to be well defined, can be found in Filus and Filus (2017b). This set of problems is, however, out of scope of this chapter, as in what follows we concentrate rather on the associated statistical analysis.

8.3 Pseudoexponential Models in Medicine

8.3.1 Application of Pseudoexponential Models to Fractal Hypothesis for Cancer

In this section, we discuss recent developments related to two areas. One is related to tumor growth and we can meet boundary problems of evolution (Nicolis, Kiselák, Porro et al. 2017); the second one is related to the structure of the tissue that can be observed from the 2-D images obtained from histology (Hermann, Mrkvička, Mattfeldt et al. 2015). In this chapter, we discuss recent papers related to both areas. When we consider fractal-based cancer diagnostics, a statistical procedure to assess the fractal dimension is often needed. We shall look for some analytical tools for discrimination between cancer and healthy ranges of fractal dimensions of tissue. Baish and Jain (2000) investigated planar tissue preparations in mice, observing the remarkably consistent scaling exponents (fractal dimensions) for tumor vasculature, even among tumor lines that have quite different vascular densities and growth characteristics. One potential problem is the reconstruction of fractal dimensions from boundaries, which may have a practical importance for Wilms tumors (Giebel 2008; Stehlík, Giebel, Prostakova et al. 2014).

On the other hand, in previous investigations, it has been shown that the texture of mammary tissue, as seen at low magnification, may be characterized quantitatively in terms of stochastic geometry (see Mattfeldt 2007; Mattfeldt, Meschenmoser, Pantle et al. 2003 and references therein). In Mrkvička and Mattfeldt (2011), the images of the mammary cases were tested for compatibility with a Boolean model (20 cases of mastopathy and 20 cases of mammary cancer, each with 10 images). In Stehlík, Mrkvička, Filus et al. (2012), we demonstrate the possibility of testing for cancer using some stochastic geometry descriptors.

The above mentioned problems are challenging and provide a good base for applications of stochastic models such as the bivariate pseudogeneralized gamma, in particular the Gumbelian type of pseudoexponential bivariate survival functions. We consider the medical trials in which the (random) sizes of tumor X_1, X_2 in two different patient populations are considered, first of all, with respect to their (in general, stochastic) dependence on a variety of circumstances. The typical circumstances are the kind of treatment, as well as some additional positive or negative stresses like a diet, smoking tobacco, or excessive use of alcohol. The reason for dependence may be the use of a common medication for both groups that changes the tumor's behavior in both compared with untreated (independent) cases. The second reason, for non-homogeneity, may be viewed as, say, *stochastic reaction* on the stresses that, possibly, only one of the groups has experienced. Notice that the two corresponding experiments may either be performed separately or together.

The data, that is, measurements of the tumor sizes taken in such medical trials from the two populations, does not, in general, show a good fit to the common bivariate normal distribution. However, the marginal X_1, X_2 samples often indicate close algebraic relationship to the exponentials. If they are dependent, the problem of such stochastic dependence modeling was, until recently, a significant problem, often only solved by forcing the real-life data to always "obey" the same bivariate normal distribution's pattern. To get around this necessity, we have chosen the bivariate pseudoexponential distribution that has been relatively recently established in the literature (Filus and Filus 2006).

Since the dependences between the two patient populations are mutual, we have chosen the first Gumbel bivariate exponential distribution that can be proven to be a separate version of the pseudoexponential class of the distributions, with the possibility of comprising a mutual physical (not merely stochastic) influence of, say, quantities X_1 and X_2. This, however old, (Gumbel) distribution (see Gumbel 1960), viewed as a type of the pseudoexponential, gains a wide range of practical interpretations that until quite recently was lacking. The existence and magnitude of the stochastic dependence is expressed by an additional parameter that we have denoted throughout by A (as the case $A=0$, which corresponds to independence), while the fact of homogeneity is described by $\mu_1 = \mu_2$.

For pseudoexponential dependence, we introduce in Filus, Filus and Stehlík (2009a, 2009b) the following assumptions:

(A1) We have two groups of planar tissues

(A2) We observe the norms $\|y_{1,i}\|, \|y_{2,i}\|, i=1,...,N$ of planar points $y_{1,i}, y_{2,i}, i=1,...,N$ from tissues 1, 2.

(A3) We formulate a log transform to obtain exponentially distributed random variables $x_{j,i} = \alpha_j \log \|y_{j,i}\|, j=1,2; i=1,...,N$, with the joint density given by the pseudoexponential model with survival function

$$P(X_1 > x_1, X_2 > x_2) = \exp(-\theta_1 x_1 - \theta_2 x_2 - \theta_2 A\phi(x_1)x_2) \qquad (8.2)$$

introduced by Filus and Filus (2007b).

We use the Pareto model for the norms of observed points (see, e.g., Filus, Filus and Stehlík 2009a,b). The Pareto model $P(\lambda, \gamma)$ is a typical model for data, with Pareto tails given by cumulative distribution function (cdf)

$$F(x) = 1 - \left(\frac{\lambda}{x}\right)^\gamma, x > \lambda$$

where:
$\gamma > 0$ is the shape parameter that characterizes the tail distribution
$\lambda > 0$ is the scale parameter.

(A4) We assume relatively a small level of dependency given by

$$\sup_{x_{1i}} |A\phi(x_{1i})| \leq 0.03 \qquad (8.3)$$

which comes from the tolerance given by Baish and Jain (2000).

This assumption can be flexibly changed for the new conditions that come from medical diagnostics. Notice that estimation of γ corresponds to the estimation of the Pareto tail, which is a complex problem (see Stehlík, Potocky, Waldl et al. 2010). If we deny dependence in the pseudoexponential model, then we may consider observations to be a sample from the gamma family that may vary maximally in scale parameter. The basic model, which is both flexible and parametric, is a generalized gamma distribution.

8.3.2 Application of Pseudoexponential Models to Dose Finding Studies

As realized during discussions after the JSM 2016 session "Development and Application of Practical and Advanced Dose-Finding Designs," the approach by Chiang and Conforti (1989) can lead to bivariate cases of pseudoexponential models, where we condition on dose exposure levels in time to one tumor. We can justify the joint survival type of model

$$P(T>t, D>d) = \exp(-\theta_1 t - \theta_2 d - \theta_2 A\phi(t)d), \quad \phi(0)=0 \quad (8.4)$$

where:
- T is the time to tumor
- D is the dose exposure of toxins
- $A>0$ is a parameter
- $\phi(t)$ is the cumulative remaining toxins in the body up to time t

Following the notation by Chiang and Conforti (1989), suppose that the individual is first exposed to toxic material amount d at age a and continue to time t. Then, in a special case of Chiang and Conforti (1989), for every t we can define the failure rate $\lambda(t,d)$ for the tumor in additive form of the mortality intensity function

$$\lambda(t,d) = -\theta_1 t - \theta_2 d - \theta_2 \frac{\delta}{v}(1-\exp(-vt))d \quad (8.5)$$

which is a special case of Equation 8.4, with A being the ratio of absorbing coefficient to discharge coefficient and the rational of the given function $\phi(t)$ lying in the remaining toxins $\int_0^t d\delta \exp(-v(t-s))ds$. The independence of the absorbing coefficient δ and the discharging coefficient of time v used by Chiang and Conforti (1989) can be rather overrealistic, and a more general version of the cumulative remaining dose can be captured by a pseudoexponential model (Equation 8.4).

8.3.3 Application of Pseudoexponential Models to Bone Mineral Density

Several locations to measure bone mineral density (BMD) are studied in the study of osteoporotic fracture (SOF) by Cummings et al. (1993) to predict hip fracture. Later on, Cummings et al. (1995) explored 16 independent risk factors by means of multivariable analysis. The multivariate relationship between these BMD values were studied by Lu et al. (2001) under multivariate normal distributional assumption. Here, we consider data from the SOF, where the following four variables (as candidates to measure BMD) are considered: STOTBMD (total spine bone mineral density, T-score) HTOTBMD (total hip BMD), NBMD (femoral neck BMD), and RDist3C (radius BMD). Figures 8.1 through 8.3 show that dependence is more sophisticated than

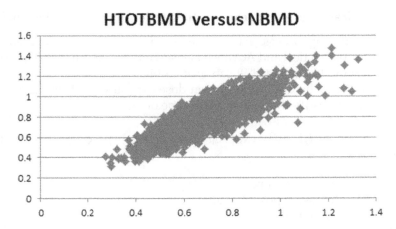

FIGURE 8.1
HTOTBMD versus NBMD.

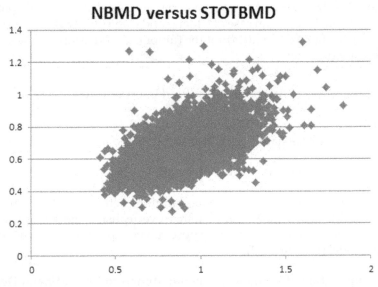

FIGURE 8.2
NMBD versus STOTBMD.

a linear relationship, thus justifying the pseudo-lognormal distribution or pseudoexponential distributions.

Realize that pseudo-lognormal distribution is a better candidate than the typically used bivariate lognormal distribution (Aitchison and Brown 1957), since it can possess much more flexible behavior of conditional expectations $E(X_2 | X_1)$ and variances $V(X_2 | X_1)$ than the standard bivariate normal or lognormal distribution. The serious disadvantage of bivariate normal and

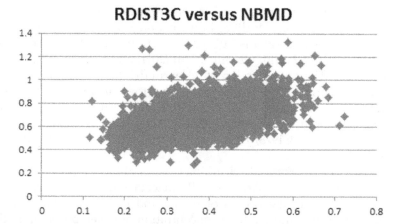

FIGURE 8.3
RDIST3C versus NBMD.

lognormal distributions is that the conditional expected value $E(X_2 | X_1 = x_1)$, which is the base for the linear regression concept, is too restricted in both cases; namely, it has a linear $a + bx_1$ or a restricted nonlinear term $a + b\log x_1$ and variance $V(X_2 | X_1)$ is constant. However, in our case, even for bivariate pseudoexponential distributions, we receive a substantially better form of

$$E(Y|X) - \mu_y = a(X_i - \mu_x) + A(X_i - \mu_x)^3.$$

8.4 Relationship of Pseudo-Weibull and Marshall–Olkin Weibull Models: Application to Administrative Arsenic Data from Parinacota, Chile

The survival function of the Morgenstein–Gumbel bivariate distribution can be obtained as

$$S(x_1, x_2) = \exp\left(-\left(x_1^m + x_2^m\right)^{\frac{1}{m}}\right)$$

Here, we can realize that for $m \to \infty$, we receive a specific form of Marshall and Olkin's (1967b) model. From another point of view, we have expansion $(x_1^m + x_2^m)^{1/m} = x_1 + x_2 + \sum_{i=1}^{\infty} \phi_i(x_1)\psi_i(x_2)$, which can be related to the pseudo-constructions. This is useful, for example, in the following concept of modeling of arsenic administrative data from Parinacota, Chile.

Arsenic and boron are chemical elements and metalloids. Arsenic is present in air, water, and land, but boron is found mainly in boron minerals, such as borax. In Chile, both arsenic and boron are present in rivers and underground waters of various localities in the country. The recommended limit for arsenic in drinking water is 0.01 mg/L, according to what is established by the WHO (World Health Organization) (2011) and the Chilean standard (NCh409, 2005). The concentration of boron in drinking water according to the WHO is 2.4 mg/L (WHO, 2011). In Europe, the standards are between 1 and 2 mg/L and in Canada it is 5 mg/L. In Chile, boron is not regulated in the regulations for drinking water (NCh409, 2005). But, it is regulated for irrigation water (NCh1333, 1978). Humans are exposed to boron mainly through food and drinking water.

In the northern zone of Chile, drinking water is obtained from sub-aquatic layers, particularly in the Arica and Parinacota region; the catchment is mainly carried out in the valleys of Azapa and Lluta, with the waters of the Lluta valley presenting a high concentration of water. boron and other heavy metals. There are studies that reveal high concentrations of arsenic and boron in drinking water in this region. For example, Tchernitchin et al. (2015) observed samples of drinking water with arsenic concentrations above 0.01 mg/L and boron concentrations of up to 14.45 mg/L. Cortes et al. (2011) took samples of drinking water from the city of Arica; the boron concentration in these samples varied between 0.22 and 11.30 mg/L.

For all of these samples, there is a need to model these extreme values of arsenic and boron present in the drinking water of the Arica and Parinacota Region.

As part of a study on drinking water quality in the Parinacota province located in the Arica and Parinacota region, the Regional Ministerial Secretary (SEREMI) of Health of Arica and Parinacota considered 28 monitoring sites. Drinking water samples were taken at each monitoring site. The samples were analyzed by the Institute of Public Health (ISP) of Chile. The analysis included the measurement of the concentration of various heavy metals such as arsenic, cadmium, chromium, mercury, and lead. We will consider the records corresponding to the arsenic concentration of each water sample coming from the monitoring sites situated in the localities shown in Table 8.1.

The recorded measurements of the arsenic concentration X in milligrams per liter in each monitoring site are not necessarily realizations of independent random variables. The joint function of two random variables U_1 and U_2 is estimated. Realizations correspond to $u_{1,i} = x_i, i = 1, \cdots, n-1$ and $u_{2,j} = x_j, j = 2, \cdots, n$. The joint distribution function F is estimated following example 6.7 presented in Marshall and Olkin (1983, p. 175), where $\bar{F} = 1 - F$ is given by $\bar{F}(\mathbf{u}) = \bar{H}_1(u_1)\bar{H}_2(u_2)\bar{H}_3(\max(u_1, u_2))$, $-\infty \leq u_1, u_2 \leq \infty$, where $\bar{H}_1(z) = \exp(-\lambda_1 z)$, $\bar{H}_2(z) = \exp(-\lambda_2 z)$, and $\bar{H}_3 = \exp(-\lambda_3 z)$ for $z > 0$ and for some λ_1, λ_2, $\lambda_3 \geq 0$ such that $\lambda_1 + \lambda_3 > 0$, $\lambda_2 + \lambda_3 > 0$. According to Galambos's example 5.5.5 (1978), the marginal distribution functions $F_1, F_2 \in D_{\min}(\Psi^*)$; then, according to proposition 4.2 (Marshall and Olkin, 1983, p. 172), $F \in D_{\min}(G)$, where

TABLE 8.1

Locality and District of Monitoring Sites

Locality	District	Locality	District
Guallatire	Putre	Putre	Putre
Socoroma	Putre	Parinacota	Putre
Caquena	Putre	Belén	Putre
Chucuyo	Putre	Saxamar	Putre
Zapahuira	Putre	Murmuntani	Putre
Lupica	Putre	Chungara	Putre
Ticnamar	Putre	Chapiquiña	Putre
Alcerreca	General Lagos	Visviri	General Lagos
Ancolacane	General Lagos	Colpitas	General Lagos
Tacora	General Lagos		

$$\bar{G}(u_1, u_2) = \exp\left\{-\left[\lambda_1 u_1 + \lambda_2 \frac{\lambda_1 + \lambda_3}{\lambda_2 + \lambda_3} u_2 + \lambda_3 \max\left(u_1, \frac{\lambda_1 + \lambda_3}{\lambda_2 + \lambda_3} u_2\right)\right]\right\}$$

$u_1, u_2 \geq 0$

Therefore, U_1, U_2 have a bivariate extreme value distribution; then, by proposition 5.1 of Marshall and Olkin (1983, p. 173), U_1, U_2 are associated, that is, they are dependent random variables. To estimate the parameters λ_1, λ_2, and λ_3, we will consider the approach suggested by Li et al. (2012, p. 2043). Therein, the survival function S is estimated based on the empirical cumulative function, that is, $S_k(u_{ik}) = 1 - S_n(u_{ik})$ (Table 8.2).

As seen in Table 8.3, $\hat{\lambda}_1 + \hat{\lambda}_3 = \hat{\lambda}_2 + \hat{\lambda}_3$; then, $\hat{\lambda}_1 = \hat{\lambda}_2$. Considering $\hat{\lambda}_1 = \hat{\lambda}_2 = 0$, we have $\hat{\lambda}_3 = 15.91007$; then, the estimate of the joint distribution $G(u_1, u_2)$ is obtained from the estimate of $\bar{G}(u_1, u_2) = \exp\{-\hat{\lambda}_3 \max(u_1, u_2)\}$ (Figure 8.4).

TABLE 8.2

Arsenic Concentration

	Min	Quartile I	Median	Quartile III	Max	Mean	St. Dev.
U_1	0.003	0.005	0.005	0.0375	0.602	0.0423	0.1154
U_2	0.003	0.005	0.005	0.0375	0.602	0.0423	0.1154

TABLE 8.3

Ordinary Least Squares Estimators

k	$\hat{\lambda}_k + \hat{\lambda}_3$	$\hat{\beta}_k$
1	15.91007	0.6399636
2	15.91007	0.6399636

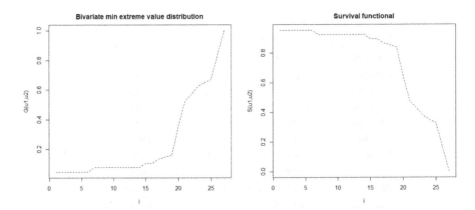

FIGURE 8.4
Graphs of bivariate distribution functions $G(u_1, u_2)$, $\bar{G}(u_1, u_2) = S(u_1, u_2)$.

8.5 Discussion by Pavlina Jordanova

For many years, scientists have looked for the most appropriate ways to model causal dependencies. It is well known that via the regression technique only, the dependence between means or quantiles can be modeled and it is not obligatory for it to be a causal dependence. For example, if we consider some sample from multivariate compound of type I, that is, the random vector, whose coordinates are defined by

$$X_{sN} = I\{N > 0\} \sum_{i=1}^{N} Y_{si}, \quad s = 1, 2, \ldots, k \tag{8.6}$$

where N is an integer valued random variable and $(Y_{1i}, Y_{2i}, \ldots, Y_{ki})$, $i = 1, 2, \ldots$ are independent identically distributed random vectors independent on N; if we then apply the regression technique to model the dependence between the coordinates of this vector, we will recognize (with difficulty) the true model (for a general theory about these distributions, see Jordanova [2016]).

The copula approach is some progress in this direction; however, it is appropriate mainly for models that can be reduced via continuous transformation to a model with uniformly distributed marginals (see Mikosch 2005). Moreover, copulas do not show the causal dependence between the coordinates in an explicit way. For example, if one considers only the copula of the joint distribution of the minima and the maxima of absolutely continuous, independent identically distributed random variables,

$$c(u,v) = \begin{cases} 0 & , \quad u \leq 0 \cup v \leq 0 \\ v & , \quad u \in (0,1], v \in (0,1], 1 - \sqrt[n]{v} > \sqrt[n]{1-u} \\ v - \left(\sqrt[n]{v} - 1 + \sqrt[n]{1-u}\right)^n & , \quad u \in (0,1], v \in (0,1], 1 - \sqrt[n]{v} \leq \sqrt[n]{1-u} \\ u & , \quad u \in (0,1], v > 1 \\ v & , \quad v \in (0,1], u > 1 \\ 1 & , \quad u > 1, v > 1 \end{cases}$$

nothing shows that it can be constructed in this way. The derivation of this form can be seen, for example, in Kostadinova and Jordanova (2009).

At the same time, from a practical point of view, the causal dependence is one of the most important forms of dependence.

It seems that the most appropriate way to model causal dependence is via decomposition of the system to its parts and presentation of the model via appropriate explicit equalities in the distribution to some investigated model. The causal dependence of the collective Cramer-Lundberg risk models on the numbers of claims of different types and the influence of the parameters on the probabilities of ruin can be seen in Jordanova et al. (2017). When we speak about the inverse-log-gamma distribution with parameters $\alpha > 0$, $\beta > 0$, with probability density function

$$P_X(x) = \begin{cases} \dfrac{\beta^\alpha}{\Gamma(\alpha)}(-\log x)^{\alpha-1} x^{\beta-1}, & x \in (0,1) \\ 0 & , \quad \text{otherwise} \end{cases}$$

nothing more than the name tells us that

$$X \stackrel{d}{=} e^{-v} \stackrel{d}{=} Y^s \stackrel{d}{=} U(0,M) \underset{M}{\wedge} U(0,1)$$

where $v \sim \text{Gamma}(\alpha, \beta)$, Y is an inverse-log-gamma distribution with parameters α, $s\beta$, $s > 0$ and the last piece of notation concerns the mixture of uniform distribution and uniform on (0, 1) distribution. This expression presents one and the same stochastic model and it is described in Jordanova, Petkova, and Stehlík (2017). The brief notation about mixtures used in the last equality seems to be very comfortable and can be seen in, for example, Johnson, Kotz and Balakrishnan (1997).

We find such examples in the correspondence between different models in many theoretical and applied stochastic models; for example, in risk theory and queueing theory, or between reliability theory and survival analysis. However, how to find the functions that describe equalities within a distribution between random elements and its members? In this area, Jerzy Filus and coauthors have made substantial recent developments on this topic and propose an interesting new model. This is the model obtained via pseudo-constructions described in the series of works of Filus, Filus, and Arnold (Filus 1990; Filus and Filus 2006, 2007a, 2008; Filus, Filus, Arnold 2010; Filus and Filus 2013, 2017b). It seems to be very promising, because it answers to some extent the question: How to insert order in discovering relations between distributions and stochastic models? It generalizes the Marshal–Olkin approach for obtaining multivariate distributions with exponential components (see, e.g., Marshall and Olkin 1967a,b or Embrechts, Lindskog, and McNeil 2001 for the description of its survival copula). It also seems to be useful in time series analysis. The Filus model starts from the most important variable for the system: T_1. It influences all other coordinates of the vector. The next important factor seems to be T_2. It influences $X_2, X_3, ...$ The next one is T_3, and so on. Following this approach, we would be able to explain the causal dependence in the system. Even if we consider only Gaussian variables $T_1, T_2, ...$ and fix the functions ϕ_i and θ_i, $i = 1, 2, ...$, we will be able to obtain relatively general classes of probability distributions and will discover interesting relations between different probability laws that may be useful in analyzing causal dependencies. Due to the generality of the model, even if the distribution of $T_1, T_2, ...$ is fixed, it would be useful to conduct some further research on some particular cases of the deterministic functions that participate in the model.

Another important topic in the papers discussed is the generalization of the Gumbel bivariate exponential distribution, defined in Gumbel (1960). In Filus, Filus and Stehlík (2009), a statistical study was conducted related to bivariate pseudoexponential distribution via its survival function, which allows us to model other multiple failures. If we concentrate only on the dependence structure of this distribution, in the continuous case we can transform the marginals to uniform and work with copulas. The bivariate pseudoexponential distribution has an interesting copula:

$$C(u_1, u_2) = \begin{cases} 0, & u_1 < 0 \text{ or } u_2 < 0 \\ u_1, & u_1 \in (0,1], u_2 > 1 \\ u_2, & u_1 > 1, u_2 \in (0,1] \\ u_1 u_2^{1 + A\phi\left(-\frac{\log u_1}{\theta_1}\right)}, & u_1 \in (0,1], u_2 \in (0,1] \\ 1, & u_1 > 1, u_2 > 1 \end{cases} \quad (8.7)$$

It generalizes the independence copula, which is obtained for $A = 0$. However, some restrictions on the functions ϕ and A that guarantee that $S(x_1, x_2)$ is a survival function should be added. For better understanding of the model, the univariate distributions would be useful. Some particular cases of the function ϕ could be an interesting topic for future research.

The consideration of survival functions is closely related to the consideration of the corresponding sum, point, and extremal processes that they generate. For example, consider the Poisson point process over unit time interval

$$N_0([0,t) \times A) = I\{N(t) > 0\} \sum_{i=1}^{N(t)} I\{(X_{i1}, X_{i2}) \in A\}, \quad t \in [0,1) \quad (8.8)$$

where (X_{11}, X_{21}), (X_{12}, X_{22}), ... are independent identically distributions with survival function $S(x,y)$ with support $[0,\infty) \times [0,\infty)$ and $N \sim \text{HPP}(\lambda)$. It is well known that

$$P(N_0([0,t) \times A) = 0) = e^{-\mu([0,t) \times A)}$$

where $\mu(\cdot)$ is the corresponding intensity measure of N_0, and if

$$(Y_1(1), Y_2(1)) = \left(\max(X_{11}, X_{12}, ..., X_{1N(1)}), \max(X_{21}, X_{22}, ..., X_{2N(1)})\right) \quad (8.9)$$

(by convention the maxima of the empty set is zero)

$$P(Y_1(1) < x_1, Y_2(1) < x_2) = P(N_0([0,1) \times A_x^c) = 0) = e^{-\mu([0,1) \times A_x^c)}$$
$$= e^{-\lambda[1 - P(X_{11} < x_1, X_{21} < x_2)]}$$

where $A_x^c = [0,\infty] \times [0,\infty) \setminus [0,x_1] \times [0,x_2]$. Therefore,

$$\mu([0,1) \times A_x^c) = \lambda\left[1 - P(X_{11} < x_1, X_{21} < x_2)\right]$$

For more about relations between sum, point, and extremal processes and their limiting behavior, see Jordanova (2005) and Jordanova and Pancheva (2004).

The parameter dependence approach is another possibility to obtain new explicit relations between distributions and, in this way, to describe causal dependence. The mixing variable usually expresses some individual behavior of the parts of the system and can be described via conditional distributions. This is discussed in the second part of this section. The mixing variables can be used to obtain new extremal processes. Such an approach is applied in Pancheva, Kolkovska, and Jordanova (2007), where we replaced the deterministic time of the extremal process with a random function. Cox (1972) carried out a similar operation with Poisson processes and Filus and Filus (2017b) generalized this approach.

Different approaches used to obtain new multivariate distributions reveal new relations between different multivariate distributions and corresponding coordinates. In this way, they provide information about new causal dependencies. Filus et al. present an application of their results in analytical cancer research and this, once again, motivates the importance of their research.

8.6 Discussion by Barry C. Arnold

Jerzy (and Lidia) recognized early the potential of the strategy of modeling bivariate distributions via one marginal distribution of, say, the first coordinate variable and the family of conditional distributions of the second variable given particular values of the first variable. In the absolutely continuous case, this approach consists of selecting a parametric family of densities for X_1 (the first coordinate variable), say $\{f_{X_1}(x_1; \underline{\theta}) : \underline{\theta} \in \Theta\}$ and combining this with a family of conditional densities $f_{X_2|X_1}(x_2 | x_1; \underline{g}(x_1))$ where $g : \mathcal{R} \to \mathcal{R}^k$ determines the parameters of the conditional densities. Higher-dimensional versions of this modeling paradigm were also described in Jerzy's work. This formulation can be visualized as one of modeling using parameter dependence. (It is regrettable that Jerzy, in his many publications, frequently fails to specifically define what he means by parameter dependence, but I believe that this description captures the spirit of Jerzy's concept.)

In reliability and survival contexts, conditioning on survival beyond a given time rather than conditioning on the time of failure is often deemed to be more appropriate. In such a setting, we might postulate a parametric model for the survival function of X_1 and then consider a parametric form for the conditional survival functions $P(X_2 > x_2 | X_1 > x_1)$ with parameters that depend on x_1. In particular, Jerzy has investigated cases in which the conditional survival function is specified in terms of the corresponding conditional hazard function (and where the marginal survival function of X_1 is also specified via its hazard function).

Models of these kinds clearly assign different roles to X_1 and X_2. It is tempting, though not absolutely necessary, to think that X_1 affects the stochastic behavior of X_2 in some way, and not conversely. Thoughts of causation inevitably arise in such scenarios, but the mathematical formulation need not concern itself with the issue of whether there is or is not a causative relationship between the variables.

One potentially undesirable feature of such parametric dependence models is that although the marginal distribution of X_1 is typically quite "nice"—indeed, it is often a member of a well-known parametric family of tractable distributions—the marginal distribution of X_2 is typically quite complicated and often decidedly "not nice."

In the survival context, Jerzy has recently considered models that treat the marginal variables in a more symmetric fashion. For example, consider Equation 8.1 in the interview in Section 8.2, which we repeat for convenience, using X, Y instead of X_1, X_2 as labels for the variables:

$$S(x,y) = P(X > x, Y > y)$$

$$= \exp\left\{-\int_0^x \lambda_1(t)\,dt - \int_0^y \lambda_2(u)\,du - \int_0^x \int_0^y \psi(t,u)\,du\,dt\right\}$$

$$= \exp\{-\Lambda_1(x) - \Lambda_2(y) - \Lambda_{12}(x,y)\}$$

where the Λ functions are defined in the obvious way (they are integrated hazard functions). If, in this expression, $\lambda_1(t)$ is the hazard function of X and $\lambda_2(u)$ is the hazard function of Y, then, as Jerzy notes, we can write

$$S(x,y) = P(X > x)P(Y > y)\exp\{-\Lambda_{12}(x,y)\}$$

It is convenient to eliminate the exponential function in this expression and to write

$$S_{X,Y}(x,y) = S_X(x)S_Y(y)D_{X,Y}(x,y) \qquad (8.10)$$

The function $D_{X,Y}(x,y)$ necessarily admits the following representation

$$D_{X,Y}(x,y) = \frac{S_{X,Y}(x,y)}{S_X(x)S_Y(y)} \qquad (8.11)$$

and can indeed be viewed, as Jerzy remarks, as a competitor of the copula in providing a link between the marginal survival functions and the joint survival function.

Observe that a distribution function version of Equation 8.11 is well defined for any pair of jointly distributed random variables (they do not have to be nonnegative). This parallel dependence function takes the form

$$\tilde{D}_{X,Y}(x,y) = \frac{F_{X,Y}(x,y)}{F_X(x)F_Y(y)} \qquad (8.12)$$

Local dependence functions defined in terms of densities (in discrete and absolutely continuous cases) have been discussed in the literature (especially in discrete cases). Such functions are of the form

$$d_{X,Y}(x,y) = \frac{f_{X,Y}(x,y)}{f_X(x)f_Y(y)} \tag{8.13}$$

I am not aware of any published discussion of the dependence functions (Equations 8.11 and 8.12). Here, once more, Jerzy has introduced a quite general concept and has left it up to us to work on specific details. He is usually off the topic by the time we see it, and is busily pursuing some other intriguing way of visualizing the possible structures of multivariate distributions.

Perhaps a fitting finale to this commentary will be to raise two characterization questions growing out of introspection centered on Jerzy's survival dependence function (Equation 8.11). Specifically, we ask two questions:

1. For two given marginal survival functions $S_X(x)$ and $S_Y(y)$, for what functions $J(x,y)$ will $S_X(x)S_Y(y)J(x,y)$ be a valid joint survival function?

2. What functions $J^*(x,y)$ will be such that $S_X(x)S_Y(y)J^*(x,y)$ is a valid joint survival function for any pair of marginal survival functions $S_X(x)$ and $S_Y(y)$?

The functions $J^*(x,y)$ identified in Equation 8.2 will be competitors to copulas in constructing bivariate models. Note that *any* bivariate survival function will be a valid example of $J^*(x,y)$. Note, also, that although $S_X(x)S_Y(y)J^*(x,y)$ will be a valid joint survival function, it will not have $S_X(x)$ and $S_Y(y)$ as its marginal survival functions.

8.7 Reply by Jerzy Filus to Barry C. Arnold's Comments

As Barry had agreed, the main formula for any bivariate survival function can be expressed in the factored form:

$$S(x,y) = S_1(x)S_2(y)J(x,y) \tag{8.14}$$

where $S_1(x), S_2(y)$ are the marginal survival functions of the random variables, say, X, Y. The *dependence function* $J(x,y)$, that we will now call the *joiner*, is uniquely determined by $S(x,y)$ by means of the obvious formula:

$$J(x,y) = S(x,y)/S_1(x)S_2(y) \tag{8.15}$$

whenever (x, y) is in the support of $S(x, y)$ otherwise $J(x, y)=0$.

At this point, recall that both functions $S_1(x), S_2(y)$ are indeed the marginals of $S(x,y)$ and are uniquely determined by the joint survival function $S(x,y)$ on substituting $S_1(x) = S(x,0)$ and

$$S_2(y) = S(0,y)$$

The first question Barry posed at end of his discussion that we will now try to answer is:

1. For two given marginal survival functions $S_1(x)$ and $S_2(y)$ (in our notation), for what functions $J(x,y)$ will $S_1(x)S_2(y)J(x,y)$ be a valid survival function?

From Equation 8.15, we obtain

$$S_2(y)J(x,y) = S(x,y)/S_1(x) = S_2(y|x) = P(Y \geq y | X \geq x) \quad (8.16)$$

Thus, to find the required conditions for $J(x,y)$, we realize that it reduces to the requirement that, for each value of x, the product $S_2(y)J(x,y)$ is a valid (univariate) survival function in y. For any x, we must then have $S_2(0)J(x,0) = 1$ and, therefore, we obtain $J(x,0) = 1$. Moreover, the product $S_2(y)J(x,y)$ is nonnegative [and so is $J(x,y)$] and a nonincreasing function of y. There is obviously no necessity that $J(x,y)$ alone is (given a fixed x) always nonincreasing when the product $S_2(y)J(x,y)$ always is.

There is also no necessity that $J(x,y) \to 0$ as $y \to \infty$.

Resuming, the sufficient and necessary conditions for $J(x,y)$ so that, given the fixed functions $S_1(x), S_2(y)$, the product $S_1(x)S_2(y)J(x,y)$ is to be a valid survival function, are:

1. $J(x,y)$ is a nonnegative real function and, for each $x, J(x,0) = 1$.
2. It is continuous from the right with respect to each of its variables.
3. For each x, the product $S_2(y)J(x,y)$ is a nonincreasing function of y.
4. Additionally, if the corresponding joint probability density for $S(x,y)$ exists, the second mixed derivative $\dfrac{\partial^2}{\partial x \partial y}(S_2(y)J(x,y)S_1(x))$

   does exist and is nonnegative. The nonnegativity condition restricts the class of the joiners $J(x,y)$, given the marginals $S_1(x), S_2(y)$. That subject is developed in more detail in Filus and Filus (2017a). More can be found in Filus and Filus (2017b).

The joiners $J(x,y)$ are much easier to find for practical applications if we replace condition 3. by the stronger condition that, for each $x, J(x,y)$ (alone) is a nonincreasing function of y.

Such functions, among others, are all functions in the form

$$J(x,y) = \exp\left[-\int_0^x\int_0^y \psi(t;u)\,du\,dt\right]$$

with $\psi(t;u) \geq 0$ satisfying the integral inequality (see Filus and Filus 2017b) with respect to the function $\psi(t;u)$ associated with condition 4. Many nice solutions for $\psi(t;u)$ [and so for $J(x,y)$] can be found easily.

As one can see, the class of the joiners proper for the pair $S_1(x), S_2(y)$ is quite wide. However, the inequality associated with condition 4. restricts it pretty strongly.

In any case, given any $S_1(x), S_2(y)$ (not necessarily from the same class of distribution), many bivariate models $S(x,y)$ having $S_1(x), S_2(y)$ as their common marginals can be constructed, although not every joiner $J(x,y)$ is proper for a given fixed pair $\{S_1(x), S_2(y)\}$. Obviously, all these considerations can also be applied to the product

$$S_1(x)J(x,y) = S(x,y)/S_2(y) = S_1(x|y) = P(X \geq x | Y \geq y)$$

The two representations of the (same) joiner are complementary.

As for the possible association of this survival approach with the copula idea, we realize the similarity to Sklar's theorem:

"Every multivariate distribution function $F(x_1,\ldots,x_n)$ can be expressed in terms of its marginals $F_i(x_i)$ $(i=1,\ldots,n)$ and the copula $C(), H$ so that $F(x_1,\ldots,x_n) = (F_1(x_1),\ldots,F_n(x_n))$."

In the case we consider, the corresponding version might be:

"Every bivariate survival function $S(x,y)$ can be expressed in terms of the product of its marginals and the joiner $J(x,y)$ that is uniquely determined by $S(x,y)$."

The similarities between these two statements are clear and left to the reader.

8.8 Rejoinder on Joiners by Barry C. Arnold

Consider the following question: For two given survival functions $S_1(x)$ and $S_2(y)$, for what functions (joiners) $J(x,y)$ will $S_1(x)S_2(y)J(x,y)$ be a valid joint survival function with S_1 and S_2 as its marginal survival functions?

Answer: Let $\mathcal{M}$ denote the class of all bivariate survival functions with marginal survival functions S_1 and S_2. For each $H(x,y) \in \mathcal{M}$, define

$$J_H(x,y) = \frac{H(x,y)}{S_1(x)S_2(y)}.$$

The class of all valid joiners (yielding joint survival functions with marginal survival functions S_1 and S_2) then consists of all functions of the form $J_H(x,y)$ where $H \in \mathcal{M}$.

Acknowledgments: The work was supported by project Fondecyt Proyecto Regular No. 1151441, Project LIT-2016-1-SEE-023, and by the bilateral projects Bulgaria–Austria, 2017–2019, contract number 01/8, 23/08/2017. We also acknowledge the help of Mag. Gerhard Holzknecht with LateX.

References

Aalen, O. O. 1989. A linear regression model for the analysis of the life times. *Statistics in Medicine* 8: 907–925.

Aitchison, J. and J. A. C. Brown. 1957. *The Lognormal Distribution, with Special Reference to Its Uses in Economics.* Cambridge, University Press.

Arnold, B. C., E. Castillo, and J. M. Sarabia. 1993. Conjugate exponential family priors for exponential family likelihoods. *Statistics* 25: 71–77.

Baish, J. W. and R. K. Jain. 2000. Fractals and cancer. *Cancer Research* 60: 3683–3688.

Chiang, C. L. and P. M. Conforti. 1989. *Mathematical Biosciences* 94: 1–29.

Cox, D. R. 1972. Regression models and life-tables. *Journal of the Royal Statistical Society. Series B (Methodological)* 34(2): 187–220.

Cortes, S., E. Reynaga-Delgado, A. M. Sancha, and C. Ferreccio. 2011. Boron exposure assessment using drinking water and urine in the north of Chile. *Science of the Total Environment* 410: 96–101.

Cummings, S. R. Black, D. M., Nevitt, M. C., Browner, W., Cauley, J., Ensrud, K., Genant, H. K., et al. 1993. Bone density at various sites for prediction of hip fractures. The Study of Osteoporotic Fractures Research Group. *Lancet.* 1993 Jan 9;341(8837):72–5.

Cummings, S. R., Nevitt, M. C., Browner, W. S., Stone, K., Fox, K. M., Ensrud, K. E., Cauley, J., et al. 1995. Risk factors for hip fracture in white women. Study of Osteoporotic Fractures Research Group. *N Engl J Med.* 1995 Mar 23;332(12):767–73.

Embrechts, P., F. Lindskog, and A. McNeil. 2001. *Modelling dependence with copulas.* Rapport technique, Département de mathématiques, Institut Fédéral de Technologie de Zurich, Zurich.

Filus, J. K. 1978. Analysis of system reliability with secondary failures (original Polish title: Analiza niezawodnosci systemow z uwzglednieniem uszkodzen wtornych), PhD thesis, Systems Research Institute of the Polish Academy of Sciences, Warsaw, Poland.

Filus, J. K. 1986. A problem in reliability optimization. *Journal of Operational Research Society* 37(4): 407–412.W

Filus, J. K. 1987. The load optimization of a repairable system with gamma-distributed time-to-failure. *Reliability Engineering* 18: 275–284.

Filus, J. K. 1990. Some modifications of the multidimensional Marshall and Olkin model. *International Journal of Systems Science* 1145–1152.

Filus, J. K. and L. Z. Filus. 2000. A class of generalized multivariate normal densities. *Pakistan Journal of Statistics* 16(1): 11–32.

Filus, J. K. and L. Z. Filus. 2001. On some bivariate pseudonormal densities. *Pakistan Journal of Statistics* 17(1): 1–9.

Filus, J. K. and L. Z. Filus. 2006. On some new classes of multivariate probability distributions. *Pakistan Journal of Statistics* 22: 21–42.

Filus, J. K. and L. Z. Filus. 2007a. On new multivariate probability distributions and stochastic processes with system reliability and maintenance applications. *Methodology and Computing in Applied Probability* 9: 425–446.

Filus, J. K. and L. Z. Filus. 2007b. On pseudonormal extension of the class of multivariate normal probability distributions. *Communications in Dependability and Quality Management* 10: 41–51.

Filus, J. K. and L. Z. Filus. 2008. On multicomponent system reliability with microshocks-microdamades type of components' interaction. *Lecture Notes in Engineering* and *Computer Science, IMECS 2008* II: 1946–1951.

Filus, J. K. and L. Z. Filus. 2013. A method for multivariate probability distributions construction via parameter dependence. *Communications in Statistics-Theory and Methods* 42.4: 716–721.

Filus, J. K. and L. Z. Filus. 2017a. Bivariate probability distributions, a universal form, *15th International Conference on Numerical Analysis and Applied Mathematics*, September 25–30, Thessaloniki, Greece, presentation.

Filus, J. K. and L. Z. Filus. 2017b. The Cox-Aalen models as framework for construction of bivariate probability distributions, universal representation. *Journal of Statistical Science and Application* 5: 56–63.

Filus, J. K. and P. Piasecki. 1980. The reliability of the system with dependent components. *Methods of Operations Research* 38: 303–313.

Filus, J. K., L. Z. Filus and Stehlík M. 2009a. Pseudoexponential modelling of cancer diagnostic testing. *Biometrie und Medizinische Informatik: Greifswalder Seminarberichte* 15: 41–54.

Filus, J. K., L. Z. Filus and M. Stehlík. 2009b. Pseudo-exponential modelling of cancer diagnostic: Testing, estimation and design. In: Kitsos, C. P. and Caroni, C. (eds), *Proceedings of ICCRA* 3: 1–16.

Filus, J. K., L. Z. Filus, and B. C. Arnold. 2010. Families of multivariate distributions involving "triangular" transformations. *Communications in Statistics: Theory and Methods* 39: 107–116.

Freund, J. E. 1961. A bivariate extension of the exponential distribution. *JASA* 56: 971–977.

Galambos, J. 1978. *The Asymptotic Theory of Extreme Order Statistics.* Wiley, New York.

Giebel, S. M. 2008. Statistical analysis of renal tumours with infants. *Bulletin de la Société des sciences médicales du Grand-Duché de Luxembourg* 1: 121–130.

Gumbel, E. J. 1960. Bivariate exponential distribution. *Journal of the American Statistical Association* 50: 698–707.

Hermann, P., T. Mrkvička, T. Mattfeldt, M. Minarova, K. Helisova, O. Nicolis, F. Wartner and M. Stehlík. 2015. Fractal and stochastic geometry inference for breast cancer: A case study with random fractal models and Quermass-interaction process. *Statistics in Medicine* 34: 2636–2661.

Johnson, N. L. and S. Kotz. 1972. *Distributions in Statistics: Continuous Multivariate Distributions.* Wiley, New York.

Johnson, N. L., S. Kotz and N. Balakrishnan. 1997. *Discrete Multivariate Distributions,* vol. 165. Wiley, New York.

Jordanova, P. K. 2005. Multivariate functional extremal criterion. PhD thesis (in Bulgarian), Bulgarian Academ y of Science, Sofia, Bulgaria. https://www.researchgate.net/publication/303486852_Multivariate_extremal_criterion_PhD_thesis

Jordanova, P. K. 2016. Multivariate compounds with equal number of summands. *Pliska Studia Mathematica* 27: 5–22.

Jordanova, P. K. and E. I. Pancheva. 2004. A functional extremal criterion. *Journal of Mathematical Sciences* 121.5: 2636–2644.

Jordanova, P. K., M. P. Petkova, and M. Stehlík. 2017. Inverse Log-Gamma-G processes. *AIP Conference Proceedings* 1895(1): 030003-1–030003-11.

Jordanova, P.K., and M. Stehlík. 2018. On multivariate modifications of Cramer–Lundberg risk model with constant intensities. Stochastic Analysis and Applications, DOI: 10.1080/07362994.2018.1471403

Kostadinova, K. and P. Jordanova. 2009. Copulas of non-monotone functions of random variables. *Proceedings with Scientific Works of Shumen University* 1: 53–60.

Kotz, S., N. Balakrishan, and N. L. Johnson. 2000. *Continuous Multivariate Distributions*. Vol. 1: Models and Applications, 2nd edn. Wiley.

Li, Y., J. Sun, and S. Song. 2012. Statistical analysis of bivariate failure time data with Marshall–Olkin Weibull models. *Computational Statistics and Data Analysis* 56: 2041–2050.

Lu, Y., H. K. Genant, J. Shepherd, S. Zhao, A. Mathur, T. P. Fuerst, and S. R. Cummings. 2001. Classification of osteoporosis based on bone mineral densities. *Journal of Bone and Mineral Research* 16(5): 901–910.

Marshall, A. W. and I. Olkin. 1967a. A generalized bivariate exponential distribution. *Journal of Applied Probability* 4.2: 291–302.

Marshall, A. W. and I. Olkin. 1967b. A multivariate exponential distribution. *Journal of the American Statistical Association* 62: 30–41.

Marshall, A. W. and I. Olkin. 1983. Domains of attraction of multivariate extreme value distributions. *The Annals of Probability* 11: 168–177.

Mattfeldt, T. 2003. Classification of binary spatial textures using stochastic geometry, nonlinear deterministic analysis and artificial neural networks. *International Journal Pattern Recognition and Artificial Intelligence* 17: 275–300.

Mattfeldt, T., D. Meschenmoser, U. Pantle, and V. Schmidt. 2007. Characterization of mammary gland tissue using joint estimators of Minkowski functionals. *Image Analysis Stereology* 26: 13–22.

Mikosch, T. 2005. Copulas: Tales and facts. *Extremes* 9(1): 3–20.

Mrkvička, T. and T. Mattfeldt. 2011. Testing histological images of mammary tissues on compatibility with the Boolean model of random sets. *Image Analysis and Stereology* 30: 11–18.

Nicolis, O., J. Kiselak, F. Porro, and M. Stehlík. 2017. Multi-fractal cancer risk assessment. *Stochastic Analysis and Applications* 35(2): 237–256.

Pancheva, E. I., E. T. Kolkovska, and P. K. Jordanova. 2007. Random time-changed extremal processes. *Theory of Probability and Its Applications* 51(4): 645–662.

Stehlík, M. 2016. On convergence of topological aggregation functions. *Fuzzy Sets and Systems* 287: 48–56

Stehlík, M., S. M. Giebel, J. Prostakova, and J. P. Schenk. 2014. Statistical inference on fractals for cancer risk assessment. *Pakistan Journal of Statistics* 30(4): 439–454.

Stehlík, M., T. Mrkvička, J. Filus, and L. Filus. 2012. Recent development on testing in cancer risk: A fractal and stochastic geometry, invited paper for *Journal of Reliability and Statistical Studies* 5: 83–95.

Stehlík, M., R. Potocký, H. Waldl, and Z. Fabián. 2010. On the favourable estimation of fitting heavy tailed data. *Computational Statistics* 25: 485–503.

Sumetkijakan, S. 2017. Non-atomic bivariate copulas and implicitly dependent random variables. *Journal of Statistical Planning and Inference* 186.

Taylor, H. M. 1979. The time to failure of fiber bundles subjected to random loads. *Advances in Applied Probability* 11(3): 527–541.

Tchernitchin, A. N., J. Ríos, I. Cortés, and L. Gaete. 2015. *Polimetales en agua de Arica-Parinacota: Posibles orígenes y efectos en salud.* XIV Congreso Geológico. La Serena, Chile.

Trutschnig, W. and J. Fernandez-Sanchez. 2014. Copulas with continuous, strictly increasing singular conditional distribution functions. *Journal of Mathematical Analysis and Applications* 410: 1014–1027.

WHO (World Health Organization). 2011. *Guidelines for Drinking Water Quality.* Geneva. World Health Organization.

9
On Reliability of Renewable Systems

Vladimir Rykov

CONTENTS

9.1 Introduction and Motivation...173
9.2 Problem Setting and Notation 175
9.3 Reliability Function Study..176
   9.3.1 Reliability Function Calculation..176
   9.3.2 Reliability Function Sensitivity Analysis.................................178
9.4 Stationary Probabilities..180
   9.4.1 Partial Repair...180
   9.4.2 Full Repair 184
9.5 Quasi-Stationary Probabilities..188
9.6 Sensitivity Analysis..191
   9.6.1 Partial Repair...191
   9.6.2 Full Repair 192
   9.6.3 Quasi-Stationary Probabilities 193
9.7 Conclusion 194
9.8 Acknowledgments..194
References..194

9.1 Introduction and Motivation

The stability of different systems characteristics on changes of initial states or exterior factors is the key problem in all natural sciences. For stochastic systems, stability often means insensitivity or low sensitivity of their output characteristics to the shapes of some input distributions. One of the earliest results concerning insensitivity of systems characteristics to the shape of service time distribution was obtained in 1957 by B. Sevast'yanov [1], who proved the insensitivity of Erlang formulas to the shape of service time distributions with fixed mean values for loss queueing systems with Poisson input flow. In 1976, I. Kovalenko [2] found the necessary and sufficient conditions for the insensitivity of stationary reliability characteristics of redundant renewable systems with exponential lifetime and general repair time distributions of its components to the shape of the latter. These conditions consist of a sufficient amount of repair facilities, that is, with the possibility of an immediate start on the repair of any failed element. The sufficiency of this condition for the case of general lifetime and repair time distributions was found by V. Rykov in 2013 [3] with the help of the multidimensional alternative processes theory. However, in the case of limited possibilities for restoration, these results do not hold, as was shown, for example, in reference [4] with the help of the additional variable method.

On the other hand, in the series of works by B. Gnedenko (1964) and A. Solov'ev (1970) [5–7], it was shown that under quick restoration, the reliability function of a cold-standby double redundant heterogeneous system tends to the exponential one for any lifetime and repair time distributions of its components. This result also indicates the asymptotic insensitivity of the reliability characteristics of such a system to the shapes of its components' lifetime and repair time distributions.

In the papers of V. Rykov with coauthors [8–10], the problem of systems steady-state reliability characteristics' sensitivity to the shape of lifetime and repair time distributions of its components for the same type of system was considered for the case when one of the input distributions (of either lifetime or repair time length) is exponential. For these models, explicit expressions for stationary probabilities have been obtained that show their evident dependence on the non-exponential distributions in the form of their Laplace–Stieltjes transforms. The numerical investigation and simulation results proposed in [11–13] demonstrate enough quick convergence of the steady-state characteristics to their limit values. This shows the practical insensitivity of the time-dependent as well as the stationary reliability characteristics to the shapes of lifetime and repair time distributions with fixed mean values. The problem of the convergence rate has not been investigated sufficiently. In Kalashnikov (1997) [14], evaluation of the convergence rate has been undertaken in terms of moments of appropriate distributions.

Some of the results represented in this chapter were announced at the International Conferences on Mathematical Methods in Reliability (MMR-2017) and Analytical and Computational Methods in Probability Theory (ACMPT-2017) and were represented in the papers without proofs [15–17]. The present chapter summarizes the previous results, generalizes some of them, and proposes new approaches for their proofs. The chapter is organized as follows. In the next section, problem setting and notation will be introduced. In Section 9.3, the reliability function is calculated in terms of Laplace transforms (LTs), and its sensitivity properties are proposed. In the following section, the steady-state probabilities (s.s.p.) of the system are examined for two types of system restoration after its failure: partial and full repair. In Section 9.5, so-called *quasi-stationary probabilities (q.s.p.)* are studied. Finally, in the last section, the asymptotic insensitivity of the system s.s.p. and q.s.p. to the shapes of their repair time distributions is considered. The chapter ends with the conclusions and the description of some problems.

9.2 Problem Setting and Notation

Consider a heterogeneous hot double redundant repairable reliability system, represented in Figure 9.1.

The following assumptions are used. The lifetimes of components are exponentially distributed random variables (r.v.) with parameters α_1 and α_2, respectively. The repair times of components have absolute continuous distributions with cumulative distribution functions (c.d.f.) $B_k(x)$ ($k=1,2$) and probability density functions (p.d.f.) $b_k(x)$ ($k=1,2$), respectively. All lifetimes and repair times are independent. The up (working) states of each component are represented by 0 and the down (failed) states by 1, respectively.

Under the assumptions considered, the system state space can be represented as $E = \{0,1,2,3\}$, with the following meanings: 0: both components are working; 1: the first component is being repaired, and the second one is working; 2: the second component is being repaired, and the first one is working; 3: both components are in the down state—the system has failed and is being repaired. Thus, the subset $E_0=\{0,1,2\}$ represents the system working states, and the subset $E_1 = \{3\}$ represents the system failure state. Also, the following terms are denote by these formulas:

- $\alpha = \alpha_1 + \alpha_2$ is the summary intensity of the system failure
- $b_k = \int_0^\infty (1-B_k(x))dx$ is the expected repair time of the kth component
- $\rho_k = \alpha_k b_k$ is the kth component availability coefficient
- $\beta_k(x) = (1-B_k(x))^{-1} b_k(x)$ is the conditional repair intensity of the kth component, given that the elapsed repair time is x;
- $\tilde{b}_k(s) = \int_0^\infty e^{-sx} b_k(x) dx$ is the LT of the repair time p.d.f. of the kth component
- $T = \inf\{t : J(t) \in E_1\}$ is the lifetime of the system.

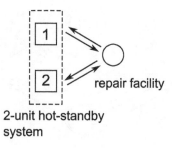

2-unit hot-standby system

FIGURE 9.1
Two-unit hot-standby repairable system with one repair facility.

In this chapter, we consider the *reliability function* of the system

$$R(t) = \mathbf{P}\{T > t\}$$

or the system lifetime distribution $F(t) = 1 - R(t)$. The system steady-state probabilities

$$\pi_j = \lim_{t \to \infty} \mathbf{P}\{J(t) = j\}$$

will also be studied. However, because no system exists infinitely long in practice, the so-called q.s.p. are more interesting characteristics,

$$\bar{\pi}_j = \lim_{t \to \infty} \mathbf{P}\{J(t) = j \mid t \leq T\}$$

which also will be investigated.

9.3 Reliability Function Study

9.3.1 Reliability Function Calculation

For the investigation of the system reliability function, the Markovization method is used. To realize it, consider the two-dimensional absorbing Markov process $Z = \{Z(t), t \geq 0\}$ with $Z(t) = (J(t), X(t))$, where $J(t)$ represents the system state and $X(t)$ is an additional variable that denotes the elapsed repair time of the $J(t)$ th component at time t. The process phase space is represented as $\mathcal{E} = \{0, (1, x), (2, x), 3\}$. This has the following meanings: 0: both components are working; $(1, x)$: the second component is working, the first one has failed and is being repaired, and its elapsed repair time is equal to x; $(2, x)$: the first component is working, the second one has failed and is being repaired, and its elapsed repair time is equal to x; 3: both components have failed and, therefore, the system has failed (absorbing state). Appropriate probabilities are denoted by $\pi_0(t)$, $\pi_1(t; x)$, $\pi_2(t; x)$, $\pi_3(t)$. The state transition graph of the system is represented in Figure 9.2.

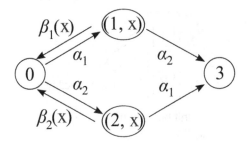

FIGURE 9.2
Absorbing system transition graph.

By the usual method of comparing the process probabilities in the closed times t and $t+\Delta$, the following Kolmogorov forward system of partial differential equations for these probabilities can be obtained

$$\frac{d}{dt}\pi_0(t) = -\alpha\pi_0(t) + \int_0^t \pi_1(t,u)\beta_1(u)du + \int_0^t \pi_2(t,u)\beta_1(u)du$$

$$\left(\frac{\partial}{\partial t} + \frac{\partial}{\partial x}\right)\pi_1(t;x) = -(\alpha_2 + \beta_1(x))\pi_1(t;x)$$

$$\left(\frac{\partial}{\partial t} + \frac{\partial}{\partial x}\right)\pi_2(t;x) = -(\alpha_1 + \beta_2(x))\pi_2(t;x)$$

$$\frac{d}{dt}\pi_3(t) = \alpha_1\int_0^t \pi_2(t;u)du + \alpha_2\int_0^t \pi_1(t;u)du \qquad (9.1)$$

jointly with the initial $\pi_0(0) = 1$ and boundary conditions

$$\pi_1(t,0) = \alpha_1\pi_0(t), \quad \pi_2(t,0) = \alpha_2\pi_0(t) \qquad (9.2)$$

The following theorem holds:

Theorem 9.1: The LTs $\tilde{\pi}(s)$ ($i \in \{0,1,2,3\}$) and $\tilde{R}(s)$ of the time-dependent probabilities $\pi_i(t)$ ($i \in \{0,1,2,3\}$) and the reliability function $R(t)$ are

$$\tilde{\pi}_0(s) = \frac{1}{s} + \psi(s)$$

$$\tilde{\pi}_1(s) = \alpha_1\frac{1-\tilde{b}_1(s+\alpha_2)}{(s+\alpha_2)(s+\psi(s))}$$

$$\tilde{\pi}_2(s) = \alpha_2\frac{1-\tilde{b}_2(s+\alpha_1)}{(s+\alpha_1)(s+\psi(s))}$$

$$\tilde{\pi}_3(s) = \frac{\alpha_1\alpha_2(\phi_1(s)+\phi_2(s))}{s(s+\alpha_1)(s+\alpha_2)(s+\psi(s))}$$

$$\tilde{R}(s) = \frac{(s+\alpha_1)(s+\alpha_2) + \alpha_1\phi_1(s) + \alpha_2\phi_2(s)}{(s+\alpha_1)(s+\alpha_2)(s+\psi(s))} \qquad (9.3)$$

where the following notation is used:

$$\phi_i(s) = (s+\alpha_i)\left(1-\tilde{b}_i(s+\alpha_{i^*})\right), (i=1,2) \tag{9.4}$$

$$\psi(s) = \alpha_1\left(1-\tilde{b}_1(s+\alpha_2)\right) + \alpha_2\left(1-\tilde{b}_2(s+\alpha_1)\right) \tag{9.5}$$

with $i^* = 2$ for $i = 1$, and $i^* = 1$ for $i = 2$.

Proof. The proof can be carried out with the help of the LT of these equations. It was demonstrated at the International Conference MMR-2017 and submitted for publication to the journal *Applied Stochastic Models in Business and Industry* (see also reference [17]). ∎

Theorem 1 allows the calculation of the time spent by the process in its transient states and mean time to failure; in particular, the following corollary holds:

Corollary 1. The mean lifetime of the system under consideration equals

$$m = \mathbf{E}[T] = \tilde{R}(0) = \frac{\alpha_1\alpha_2 + \alpha_1\left(1-\tilde{b}_1(\alpha_2)\right)\alpha_2\left(1-\tilde{b}_2(\alpha_1)\right)}{\alpha_1\alpha_2\left[\alpha_1\left(1-\tilde{b}_1(\alpha_2)\right)+\alpha_2\left(1-\tilde{b}_2(\alpha_1)\right)\right]} \tag{9.6}$$

9.3.2 Reliability Function Sensitivity Analysis

These formulas demonstrate an evident dependence of the reliability function on the shape of the repair time distribution. It is expressed in the form of the Laplace–Stieltjes transformation of the repair time distribution at the points of failure items' intensities. Of course, in the Markov case, with exponential repair time distribution, these dependencies are expressed in terms of only one parameter of the exponential distribution. From another perspective, as mentioned in the introduction, in the case of quick restoration the reliability function tends to exponential form for any repair time distribution. In this section, we consider the behavior of the reliability function instead of *quick* restoration under *rare* failures. For the model considered, rare failures means the slow intensity of failures with respect to fixed repair times. Thus, we will suppose that

$$q = \max\{\alpha_1, \alpha_2\} \to 0$$

Naturally, the asymptotic analysis should be done with respect to some scale parameter. The asymptotic mean lifetime value will be considered as such a parameter. In accordance with Equation 9.4, taking into account that under $q \to 0$ with $b_i = \mathbf{E}[B_i] = -\tilde{b}'_i(0) = \int_0^\infty (1-B_i(x))dx$ and $\rho_i = \alpha_i b_i$ one can find

$$\phi_i(0) = \alpha_i^2\left(1 - \tilde{b}_i(\alpha_{i^*})\right) \approx \alpha_i^2 b_i \alpha_{i^*} = \alpha_i \alpha_{i^*} \rho_i$$

which takes the value

$$m = \mathbf{E}[T] = \tilde{R}(0) \approx \frac{\alpha_1 \alpha_2 + \alpha_1 \phi_1(0) + \alpha_2 \phi_2(0)}{\alpha_1 \alpha_2 \psi(0)}$$

$$= \frac{1 + \rho_1 + \rho_2}{\alpha_1 \alpha_2 (b_1 + b_2)}$$

Theorem 9.2: The reliability function of the model considered in the scale $m = \mathbf{E}[T]$ has exponential form

$$\lim_{q \to 0} \mathbf{P}\left\{\frac{T}{m} > t\right\} = e^{-t}$$

Proof. The proof of the theorem can also be found in [17]. However, it is not too long to repeat here. For simplicity, instead of the large parameter m, consider the small parameter $\gamma = m^{-1}$. The asymptotic value of the normalized reliability function

$$R\left(\frac{t}{\gamma}\right) = \mathbf{P}\{\gamma T > t\}$$

under $\gamma \to 0$ can be considered in terms of its LT $\gamma \tilde{R}(\gamma s)$:

$$\gamma \tilde{R}(\gamma s) = \gamma \frac{(\gamma s + \alpha_1)(\gamma s + \alpha_2) + \alpha_1 \phi_1(\gamma s) + \alpha_2 \phi_2(\gamma s)}{(\gamma s + \alpha_1)(\gamma s + \alpha_2)(\gamma s + \psi(\gamma s))}$$

$$\approx \gamma \frac{1 + \rho_1 + \rho_2}{\gamma s(1 + \rho_1 + \rho_2) + \alpha_1 \alpha_2 (b_1 + b_2)} = \gamma \frac{1}{\gamma s + \gamma} \quad (9.7)$$

$$= \frac{1}{s + 1}$$

From here, it follows that under $\gamma \to 0$,

$$\mathbf{P}\{\gamma T > t\} = R\left(\frac{t}{\gamma}\right) \to e^{-t}$$

which concludes the proof. ∎

9.4 Stationary Probabilities

To calculate the s.s.p., we will use the same two-dimensional Markov process $Z = \{Z(t), t \geq 0\}$, with $Z(t) = (J(t), X(t))$ without the absorbing state. However, for the study of system s.s.p., we need to determine system behavior after its failure. There are at least two possibilities:

- Partial repair, when after failure the system continues to work in the same regime; that is, the repaired element continues to be repaired and, after its renewal, the system goes to state 1 or 2, depending on what type of component is repaired in state 3. Therefore, we need to divide this state into two states $(3, i)$, $(i = 1, 2)$, which means that both elements fail and the ith one is repaired.
- Full repair, when after a system failure the renewal of the whole system begins, which demands some random time with, say, c.d.f. $B_3(t)$; after this time, the system goes to state 0.

Therefore, for these two cases we need to consider different system state spaces. For the full repair case, it should be $E = \{0, 1, 2, 3\}$, while for the partial repair case, the system state should be $E = \{0, 1, 2, (3, 1), (3, 2)\}$. The system states subset $E_0 = \{0, 1, 2\}$ represents its working (up) states of the system, and the subset $E_1 = \{3\}$ or $E_1 = \{(3, 1), (3, 2)\}$ represents the system failure (down) state.

9.4.1 Partial Repair

The process Z phase space in the partial repair* case equals to $\mathcal{E} = \{0, (1, x), (2, x), ((3, 1), x), ((3, 2), x)\}$. This has the following meanings: 0: both components of the system are working; (i, x): the ith component $(i = 1, 2)$ has failed and is repaired—its elapsed repair time is equal to x—and the other component is working; $((3, i), x)$: both elements have failed, and the ith one is repaired with an elapsed time equal to x. Later, we will refer to these states as the *micro-states* of the process, while the states E are called the *macro-states* of the system and appropriate process. The transition graph is shown in Figure 9.3.

Appropriate probabilities are denoted by

$$\pi_0(t), \pi_1(t; x), \pi_2(t; x), \pi_{(3,1)}(t; x), \pi_{(3,2)}(t; x)$$

Because state 0 represents a positive atom for process Z, it is a Harris process and, therefore, the limiting probabilities exist for it and coincide with the stationary ones.

$$\pi_0 = \lim_{t \to \infty} \pi_0(t), \quad \pi_i(x) = \lim_{t \to \infty} \pi_i(t; x), \quad (i \in \{1, 2, (3, 1), (3, 2)\}$$

---

* Note that in this case the distributions to the first failure and between failures are different.

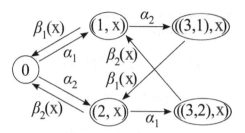

FIGURE 9.3
Transition graph of the system with partial repair.

Analogously to the case of Markov processes with discrete states spaces, the transition graph, shown in Figure 9.3, allows us to write out the following system of balance equations for the s.s.p. of the model:

- For state 0

$$\alpha_1\pi_0 = \int_0^\infty \pi_1(x)\beta_1(x)dx, \quad \alpha_2\pi_0 = \int_0^\infty \pi_2(x)\beta_2(x)dx \qquad (9.8)$$

- For states $(1,x),(2,x)$

$$\frac{d}{dx}\pi_1(x) = -(\alpha_2+\beta_1(x))\pi_1(x)$$

$$\pi_1(0) = \alpha_1\pi_0 + \int_0^\infty \pi_{(3,2)}(u)\beta_2(u)du$$

$$\frac{d}{dx}\pi_2(x) = -(\alpha_1+\beta_2(x))\pi_2(x)$$

$$\pi_2(0) = \alpha_2\pi_0 + \int_0^\infty \pi_{(3,1)}(u)\beta_1(u)du \qquad (9.9)$$

- For states $((3,1),x),((3,2),x)$

$$\frac{d}{dx}\pi_{(3,1)}(x) = -\beta_1(x)\pi_{(3,1)}(x)+\alpha_2\pi_1(x)$$

$$\frac{d}{dx}\pi_{(3,2)}(x) = -\beta_2(x)\pi_{(3,2)}(x)+\alpha_2\pi_1(x) \qquad (9.10)$$

Theorem 9.3: The micro-state s.s.p. of the system with partial repair have the form

$$\pi_1(x) = \alpha_1 \tilde{b}_1^{-1}(\alpha_2) e^{-\alpha_2 x} (1 - B_1(x)) \pi_0$$

$$\pi_2(x) = \alpha_2 \tilde{b}_2^{-1}(\alpha_1) e^{-\alpha_1 x} (1 - B_2(x)) \pi_0$$

$$\pi_{(3,1)}(x) = \alpha_1 \tilde{b}_1^{-1}(\alpha_2)(1 - e^{-\alpha_2 x})(1 - B_1(x)) \pi_0$$

$$\pi_{(3,2)}(x) = \alpha_2 \tilde{b}_2^{-1}(\alpha_1)(1 - e^{-\alpha_1 x})(1 - B_2(x)) \pi_0 \tag{9.11}$$

and the appropriate macro-state s.s.p. are

$$\pi_1 = \frac{\alpha_1(1 - \tilde{b}_1(\alpha_2))}{\alpha_2 \tilde{b}_1(\alpha_2)} \pi_0$$

$$\pi_2 = \frac{\alpha_2(1 - \tilde{b}_2(\alpha_1))}{\alpha_1 \tilde{b}_2(\alpha_1)} \pi_0$$

$$\pi_{(3,1)} = \frac{\alpha_1(1 - \tilde{b}_1(\alpha_2))(\alpha_2 b_1 - 1)}{\alpha_2 \tilde{b}_1(\alpha_2)} \pi_0$$

$$\pi_{(3,2)} = \frac{\alpha_2(1 - \tilde{b}_2(\alpha_1))(\alpha_1 b_2 - 1)}{\alpha_1 \tilde{b}_2(\alpha_1)} \pi_0 \tag{9.12}$$

where

$$\pi_0 = \left[1 + \rho_1 \frac{1 - \tilde{b}_1(\alpha_2)}{\tilde{b}_1(\alpha_2)} + \rho_2 \frac{1 - \tilde{b}_2(\alpha_1)}{\tilde{b}_2(\alpha_1)} \right]^{-1} \tag{9.13}$$

Proof. The first and the third equations of the system (9.9) are equations with separable variables, and their general solutions are

$$\pi_1(x) = C_1 e^{-\alpha_2 x}(1 - B_1(x))$$

$$\pi_2(x) = C_2 e^{-\alpha_1 x}(1 - B_2(x)) \tag{9.14}$$

The homogeneous parts of Equations (9.10) also have solutions of the forms

$$\pi_{(3,1)}(x) = C_{(3,1)}(1 - B_1(x))$$

and application of the constant variation method gives the following general solution:

$$\pi_{(3,1)}(x) = C_1\left[\left(1-e^{-\alpha_2 x}\right) + \bar{C}_{(3,1)}\right]\left(1-B_1(x)\right)$$

$$\pi_{(3,2)}(x) = C_2\left[\left(1-e^{-\alpha_1 x}\right) + \bar{C}_{(3,2)}\right]\left(1-B_2(x)\right)$$

To find the unknown constants $\bar{C}_{(3,1)}, \bar{C}_{(3,2)}$, note that the functions $\pi_{(3,1)}(x)$ and $\pi_{(3,2)}(x)$ are the probability density functions and, according to the transition graph of the process, represented in Figure 9.3, the process never visits the states $((3,1),0)$ and $((3,2),0)$, and therefore $\pi_{(3,1)}(0) \equiv \bar{C}_{(3,1)} = \pi_{(3,2)}(0) \equiv \bar{C}_{(3,2)} = 0$. Thus, finally, the expressions for probability densities $\pi_{(3,1)}(x)$ and $\pi_{(3,2)}(x)$ are

$$\pi_{(3,1)}(x) = C_1\left(1-e^{-\alpha_2 x}\right)\left(1-B_1(x)\right)$$

$$\pi_{(3,2)}(x) = C_2\left(1-e^{-\alpha_1 x}\right)\left(1-B_2(x)\right) \tag{9.15}$$

Now, from Equation 9.8, one can find that

$$\alpha_1 \pi_0 = C_1 \tilde{b}_1(\alpha_2), \quad \alpha_2 \pi_0 = C_2 \tilde{b}_2(\alpha_1)$$

Thus, constants C_1, C_2 can be represented in terms of π_0 as

$$C_1 = \alpha_1 \tilde{b}_1^{-1}(\alpha_2)\pi_0, \quad C_2 = \alpha_1 \tilde{b}_1^{-1}(\alpha_2)\pi_0. \tag{9.16}$$

The expressions (9.14, 9.15) for the so-called micro-state s.s.p. allows us, by integrating them over x, to calculate the appropriate macro-state s.s.p. Moreover, using this representation of constants C_1 and C_2 in terms of π_0 from Equation 9.16, one can represent the macro-state s.s.p. in terms of π_0 as follows:

$$\pi_1 = \int_0^\infty \pi_1(x)dx = \frac{\alpha_1\left(1-\tilde{b}_1(\alpha_2)\right)}{\alpha_2 \tilde{b}_1(\alpha_2)}\pi_0$$

$$\pi_2 = \int_0^\infty \pi_2(x)dx = \frac{\alpha_2\left(1-\tilde{b}_2(\alpha_1)\right)}{\alpha_1 \tilde{b}_2(\alpha_1)}\pi_0$$

$$\pi_{(3,1)} = \int_0^\infty \pi_{(3,1)}(x)dx = \frac{\alpha_1\left(1-\tilde{b}_1(\alpha_2)\right)(\alpha_2 b_1 - 1)}{\alpha_2 \tilde{b}_1(\alpha_2)}\pi_0$$

$$\pi_{(3,2)} = \int_0^\infty \pi_{(3,2)}(x)dx = \frac{\alpha_2\left(1-\tilde{b}_2(\alpha_1)\right)(\alpha_1 b_2 - 1)}{\alpha_1 \tilde{b}_2(\alpha_1)} \pi_0 \qquad (9.17)$$

At the least, the probability π_0 is determined from the normalizing condition

$$\pi_0 + \pi_1 + \pi_2 + \pi_{(3,1)} + \pi_{(3,2)} = 1$$

Taking into account that

$$\pi_i + \pi_{(3,i)} = \frac{\rho_i\left(1-\tilde{b}_i(\alpha_{i*})\right)}{\tilde{b}_i(\alpha_{i*})} \pi_0, \quad (i=1,2)$$

this gives the following expression

$$\pi_0 = \left[1 + \rho_1 \frac{1-r\tilde{b}_1(\alpha_2)}{\tilde{b}_1(\alpha_2)} + \rho_2 \frac{1-\tilde{b}_2(\alpha_1)}{\tilde{b}_2(\alpha_1)}\right]^{-1}$$

that proves the theorem. ∎

Corollary 2. For a homogeneous system, denote the macro-state probabilities by $\bar{\pi}_0, \bar{\pi}_1, \bar{\pi}_2$ and set $\rho = \alpha b$. The s.s.p. for homogeneous system equal:

$$\bar{\pi}_0 = \frac{\tilde{b}(\alpha)}{2\rho + \tilde{b}(\alpha)}$$

$$\bar{\pi}_1 = 2\frac{1-\tilde{b}(\alpha)}{2\rho + \tilde{b}(\alpha)}$$

$$\bar{\pi}_2 = 2\frac{\rho + 1 - \tilde{b}(\alpha)}{2\rho + \tilde{b}(\alpha)} \qquad (9.18)$$

Proof. In the homogeneous case where $\bar{\pi}_0 = \pi_0$, the macro-states probabilities $\bar{\pi}_1, \bar{\pi}_2$ can be obtained by summing the appropriate probabilities $\bar{\pi}_1 = \pi_1 + \pi_2, \bar{\pi}_2 = \pi_{(3,1)} + \pi_{(3,2)}$. Thus, by summing and substitution of appropriate values, this gives the result of the corollary.

9.4.2 Full Repair

In the case of full restoration of the system after failure, suppose that the system is fully repaired during some random time, say B_3, and passes from state 3 to state 0, which means that it becomes as new. Suppose also that the r.v. B_3 has an absolutely continuous c.d.f, $B_3(x)$, with mean value

$b_3 = \mathbf{E}[B_3] = \int_0^\infty (1 - B_3(x))dx$, appropriate p.d.f. $b_3(x)$, and transition intensity (conditional p.d.f., given that the elapsed summary system repair time equals x) equal to $\beta_3(x)$.

Regarding system investigation, consider the same two-dimensional Markov process as before, $Z = \{(J(t), X(t)), t \geq 0\}$, with phase space $\varepsilon = \{0, (1,x), (2,x), (3,x)\}$; the difference from the previous case is that in this case it contains only one state $(3,x)$, for which both elements have failed, and the system elapsed repair time is equal to x. The process transition graph is represented by Figure 9.4.

Appropriate probabilities are denoted by

$$\pi_0(t), \pi_1(t;x), \pi_2(t;x), \pi_3(t;x)$$

As in the previous case, the process Z is a positive recurrent (or Harris) one (the state 0 is its positive atom), and therefore it has limiting probabilities when $t \to \infty$, which coincide with the stationary ones

$$\pi_0 = \lim_{t \to \infty} \pi_0(t), \quad \pi_i(x) = \lim_{t \to \infty} \pi_i(t;x) \quad (i \in \{1,2,3\})$$

Using the transition graph, represented in Figure 9.4, analogously to the case of Markov processes with discrete-state spaces, one can write down the following system of balance equations for the s.s.p. of the process.

- For state 0

$$\alpha \pi_0 = \int_0^\infty \pi_1(x)\beta_1(x)dx + \int_0^\infty \pi_2(xu)\beta_1(x)dx + \int_0^\infty \pi_3(x)\beta_3(x)dx \qquad (9.19)$$

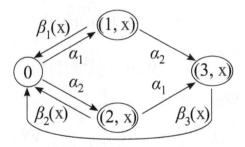

FIGURE 9.4
Transition graph of the system with full repair.

- For states $(1, x), (2, x)$

$$\frac{d}{dx}\pi_1(x) = -(\alpha_2 + \beta_1(x))\pi_1(x), \quad \pi_1(0) = \alpha_1\pi_0$$

$$\frac{d}{dx}\pi_2(x) = -(\alpha_1 + \beta_2(x))\pi_2(x), \quad \pi_2(0) = \alpha_2\pi_0 \quad (9.20)$$

- For state $(3, x)$

$$\frac{d}{dx}\pi_3(x) = -\beta_3(x)\pi_3(t;x)$$

$$\pi_3(0) = \alpha_1\int_0^\infty \pi_2(x)dx + \alpha_2\int_0^\infty \pi_1(t,x)du \quad (9.21)$$

Theorem 9.4: The stationary micro-state probabilities of the system under full repair have the form

$$\pi_1(x) = \alpha_1 e^{-\alpha_2 x}(1 - B_1(x))\pi_0$$

$$\pi_2(x) = \alpha_2 e^{-\alpha_1 x}(1 - B_2(x))o\pi_0$$

$$\pi_3(x) = \left[\alpha_1(1 - \tilde{b}_1(\alpha_2)) + \alpha_2(1 - \tilde{b}_1(\alpha_2))\right](1 - B_3(x))\pi_0 \quad (9.22)$$

with appropriate macro-states probabilities

$$\pi_1 = \frac{\alpha_1}{\alpha_2}(1 - \tilde{b}_1(\alpha_2))\pi_0$$

$$\pi_2 = \frac{\alpha_2}{\alpha_1}(1 - \tilde{b}_2(\alpha_1))\pi_0$$

$$\pi_3 = \left[\alpha_1(1 - \tilde{b}_1(\alpha_2)) + \alpha_2(1 - \tilde{b}_1(\alpha_2))\right]b_3\pi_0 \quad (9.23)$$

where π_0 is given by

$$\pi_0 = \left[1 + (1 - \tilde{b}_1(\alpha_2))\left(\frac{\alpha_1}{\alpha_2} + \alpha_2 b_3\right) + (1 - \tilde{b}_2(\alpha_1))\left(\frac{\alpha_2}{\alpha_1} + \alpha_1 b_3\right)\right]^{-1} \quad (9.24)$$

Proof. Solutions of the first parts of Equations (9.20, 9.21) are

$$\pi_1(x) = C_1 e^{-\alpha_2 x}(1 - B_1(x))$$

$$\pi_2(x) = C_2 e^{-\alpha_1 x}(1 - B_2(x))$$

$$\pi_3(x) = C_3(1 - B_3(x)) \tag{9.25}$$

Using the second parts of these equations to find the unknown constants C_i gives

$$C_1 = \alpha_1 \pi_0, \quad C_2 = \alpha_2 \pi_0, \quad C_3 = \left[\alpha_1(1-\tilde{b}_1(\alpha_2)) + \alpha_2(1-\tilde{b}_2(\alpha_1))\right]\pi_0$$

The simple integration of Equations (9.25) with respect to x allows us to find the appropriate stationary macro-state probabilities:

$$\pi_1 = \frac{\alpha_1}{\alpha_2}(1 - \tilde{b}_1(\alpha_2))\pi_0$$

$$\pi_2 = \frac{\alpha_2}{\alpha_1}(1 - \tilde{b}_2(\alpha_1))\pi_0$$

$$\pi_3 = \left[\alpha_1(1 - \tilde{b}_1(\alpha_2)) + \alpha_2(1 - \tilde{b}_1(\alpha_2))\right]b_3 \pi_0 \tag{9.26}$$

The normalizing conditions give:

$$1 = \pi_0 + \pi_1 + \pi_2 + \pi_3$$

$$= \left[1 + \frac{\alpha_1}{\alpha_2}(1 - \tilde{b}_1(\alpha_2)) + \frac{\alpha_2}{\alpha_1}(1 - \tilde{b}_2(\alpha_1))\right.$$

$$\left. + \left(\alpha_1(1 - \tilde{b}_1(\alpha_2)) + \alpha_2(1 - \tilde{b}_2(\alpha_1))\right)b_3\right]\pi_0$$

from which Equation 9.24 follows, which proves the theorem. ∎

Remark 1. For a homogeneous system, when $\alpha_i = \alpha, b_i(x) = b(x)$ for $i = 1, 2$, the last expression takes the form

$$\pi_0 = \frac{1}{1 + 2(1 - \tilde{b}(\alpha))(1 + \alpha b_3)}$$

9.5 Quasi-Stationary Probabilities

When studying the system's behavior in its life cycle (during its lifetime), rather than its stationary probabilities, the so-called q.s.p. are more interesting. They are defined as the limits of the conditional probabilities to be in any state, under the condition that the system has not failed yet:

$$\hat{\pi}_i = \lim_{t \to \infty} P\{J(t) = i \mid t \leq T\}$$

$$= \lim_{t \to \infty} \frac{P\{J(t) = i, t \leq T\}}{P\{t \leq T\}} = \lim_{t \to \infty} \frac{\pi_i(t)}{R(t)} \qquad (9.27)$$

To calculate these limits, it is possible to use the LT of the appropriate function, represented in Theorem 1 by Equations (9.3).

Theorem 9.5: The q.s.p. of the model under consideration have the form

$$\hat{\pi}_i = \lim_{t \to \infty} \frac{\pi_i(t)}{R(t)} = \frac{A_i}{A_R}, \qquad (9.28)$$

where the values A_i, A_R are the residuals of the functions $\tilde{\pi}_i(s)$ and $\tilde{R}(s)$ at the point $-\gamma$, which is the maximal root of the equation

$$\psi(s) = -s \qquad (9.29)$$

Proof. For direct calculation of the q.s.p., we require the inversion of the LTs of the appropriate functions. This is not a simple problem in the general case. Nevertheless, to calculate the q.s.p. we use their LTs directly. Note that the behavior of the functions $\pi_i(t)$ and $R(t)$ for $t \to \infty$ depends on the roots of their LT denominators. Note, now, that the denominators of these functions' LTs are almost the same, and the behavior of the functions $\pi_i(t)$ and $R(t)$ for $t \to \infty$ depends mostly on the maximal root (with minimal absolute value). The denominator of the function $\tilde{R}(s)$ has only negative roots: $s_1 = -\alpha_1, s_2 = -\alpha_2$ and the root of the equation

$$\psi(s) = -s \qquad (9.30)$$

Let us show that the last root is negative and minimal in absolute value. We will consider the solution of this equation for real values of s and suppose, for certainty, that $-\alpha_2 < -\alpha_1$. For real values of s, the functions $\tilde{b}_i(s + \alpha_i)$ are quite monotone ones (see reference [18], vol. 2) and are therefore convex; thus, the functions $1-\tilde{b}_i(\cdot)$ are concave, as well as their linear combination, which is the function $\psi(s)$. This shows that Equation 9.30 has a unique root $s = -\gamma$, which satisfies the inequality $-\alpha_1 < -\gamma < 0$ (see Figure 9.5).

This argument shows that the functions $\pi_i(t)$ and $R(t)$ have the forms

$$\pi_0(t) = A_0 e^{-\gamma t}(1 + \epsilon_0(t))$$

$$\pi_{(1,2)} \equiv \pi_1(t) + \pi_2(t) = A_{(1,2)} e^{-\gamma t}(1 + \epsilon_{(1,2)}(t))$$

$$\pi_3(t) = A_3 e^{-\gamma t}(1 + \epsilon_3(t))$$

$$R(t) = A_R e^{-\gamma t}(1 + \epsilon_R(t))$$

where functions $\epsilon_i(t)$ $(i = 1,(1,2),3)$ and $\epsilon_R(t)$ are infinitely small.

These representations allow us to calculate q.s.p. (9.27) as follows:

$$\hat{\pi}_i = \lim_{t \to \infty} \frac{\pi_i(t)}{R(t)} = \frac{A_i}{A_R} = \lim_{s \to -\gamma} \frac{\tilde{\pi}_i(s)}{\tilde{R}(s)} \quad (9.31)$$

where the values A_i, A_R are the residuals of the functions $\tilde{\pi}_i(s)$ and $\tilde{R}(s)$ at the point $-\gamma$.

In a special homogeneous case, when $\alpha_1 = \alpha_2 = \alpha$ and $\tilde{b}_1(s) = \tilde{b}_2(s) = \tilde{b}(s)$ the function $\psi(s)$ takes the value

$$\psi(s) = 2\alpha(1 - \tilde{b}(s + \alpha))$$

the investigation can be simplified. ∎

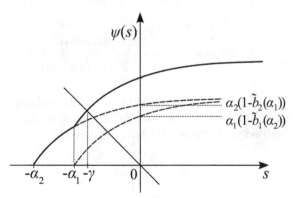

FIGURE 9.5
Root of the equation $\psi(s) = -s$.

Taking into account the expressions (9.3) for the LTs $\tilde{\pi}(s)$ and $\tilde{R}(s)$ of probabilities $\pi_i(t)$ ($i = 0,1,2,3$) and of reliability function $R(t)$, this reasoning, after some calculations, allows us to prove the following theorem.

Corollary 3. The q.s.p. of the model under consideration have the form

$$\hat{\pi}_0 = \lim_{t\to\infty} \frac{\pi_0(t)}{R(t)} = \lim_{s\to -\gamma} \frac{\tilde{\pi}_0(s)}{\tilde{R}(s)}$$

$$= \left[1 + \frac{\alpha_1}{\alpha_2 - \gamma}\left(1 - \tilde{b}_1(\alpha_2 - \gamma)\right) + \frac{\alpha_2}{\alpha_1 - \gamma}\left(1 - \tilde{b}_2(\alpha_1 - \gamma)\right)\right]^{-1}$$

$$\hat{\pi}_1 = \lim_{t\to\infty} \frac{\pi_1(t)}{R(t)} = \lim_{s\to -\gamma} \frac{\tilde{\pi}_1(s)}{\tilde{R}(s)}$$

$$= \alpha_1 \frac{\left(1 - \tilde{b}_1(\alpha_2 - \gamma)\right)(\alpha_1 - \gamma)}{(\alpha_1 - \gamma)(\alpha_2 - \gamma) + \alpha_1\phi_1(-\gamma) + \alpha_2\phi_2(-\gamma)} \qquad (9.32)$$

$$\hat{\pi}_2 = \lim_{t\to\infty} \frac{\pi_2(t)}{R(t)} = \lim_{s\to -\gamma} \frac{\tilde{\pi}_2(s)}{\tilde{R}(s)}$$

$$= \alpha_2 \frac{\left(1 - \tilde{b}_2(\alpha_1 - \gamma)\right)(\alpha_2 - \gamma)}{(\alpha_1 - \gamma)(\alpha_2 - \gamma) + \alpha_1\phi_1(-\gamma) + \alpha_2\phi_2(-\gamma)}$$

Thus, for example, for the homogeneous case one may obtain

$$\hat{\pi}_0 = \lim_{s\to -\gamma} \frac{\tilde{\pi}_0(s)}{\tilde{R}(s)} = \frac{\alpha - \gamma}{\alpha - \gamma + 2\alpha\left(1 - \tilde{b}(\alpha - \gamma)\right)}$$

$$\hat{\pi}_{(1,2)} = \hat{\pi}_1 + \hat{\pi}_2 = \frac{2\alpha\left(1 - \tilde{b}(\alpha - \gamma)\right)}{\alpha - \gamma + 2\alpha\left(1 - \tilde{b}(\alpha - \gamma)\right)}$$

Moreover, for the Markov case, when $b(t) = \beta e^{-\beta t}$, after substitution of appropriate expressions for $\tilde{b}(s + \alpha) = \beta(s + \alpha + \beta)^{-1}$, these formulas coincide with expressions obtained with the usual Markov approach:

$$\hat{\pi}_0 = \frac{\alpha + \beta - \gamma}{3\alpha + \beta - \gamma}$$

$$\hat{\pi}_1 + \hat{\pi}_2 = \frac{2\alpha}{3\alpha + \beta - \gamma}$$

9.6 Sensitivity Analysis

All these results show the evident sensitivity of the considered systems to the shapes of their components' repair time distributions. However, under rare failures, this sensitivity becomes negligible for all reliability characteristics of the systems under consideration. For the reliability function, this was shown in Section 9.3.2 (see also reference [15]). In this section, the asymptotic insensitivity of the s.s.p. and q.s.p. of the systems considered to the shapes of their components' repair time distributions under rare failures will be shown.

9.6.1 Partial Repair

The s.s.p. of the system under partial failures are represented by Equations 9.11 and 9.13. With the help of the Taylor expansion for $q = \max\{\alpha_1, \alpha_2\} \to 0$, the following theorem can be proved.

Theorem 9.6: Under components' rare failures, the s.s.p. of the system considered with partial repair take the form

$$\pi_0 \approx \frac{1}{1+\rho_1\alpha_2 b_1 + \rho_2\alpha_1 b_2}$$

$$\pi_i \approx \frac{\rho_i}{1+\rho_1\alpha_2 b_1 + \rho_2\alpha_1 b_2} \quad (i=1,2)$$

$$\pi_{(3,i)} \approx \frac{\rho_i(\alpha_{i^*} b_1 - 1)}{1+\rho_1\alpha_2 b_1 + \rho_2\alpha_1 b_2} \quad (i=1,2) \tag{9.33}$$

which show their asymptotic insensitivity to the shapes of their components' repair distributions.

Proof. Expansion around zero of $q = \max\{\alpha_1, \alpha_2\}$, the formula for π_0 in Equation 9.13, gives

$$\pi_0 = \left[1+\rho_1\frac{1-\tilde{b}_1(\alpha_2)}{\tilde{b}_1(\alpha_2)} + \rho_2\frac{1-\tilde{b}_2(\alpha_1)}{\tilde{b}_2(\alpha_1)}\right]^{-1}$$

$$\approx [1+\rho_1\alpha_2 b_1 + \rho_2\alpha_1 b_2]^{-1}$$

Analogously, the expansion around zero of $q = \max\{\alpha_1, \alpha_2\}$ in the formulas for π_i in (9.17) gives

$$\pi_i = \frac{\alpha_i\left(1-\tilde{b}_i(\alpha_{i^*})\right)}{\alpha_{i^*}\tilde{b}_1(\alpha_{i^*})}\pi_0 \approx \rho_i\pi_0$$

while for $\pi_{(3,i)}$, the same procedure gives

$$\pi_{(3,i)} = \frac{\alpha_i(1-\tilde{b}_i(\alpha_{i^*}))(\alpha_{i^*}b_i-1)}{\alpha_{i^*}\tilde{b}_i(\alpha_{i^*})}\pi_0 \approx \rho_i(\alpha_{i^*}b_i-1)\pi_0$$

Taking into account an appropriate expression for π_0, one can obtain the results of the theorem.

9.6.2 Full Repair

Analogous to the previous case, the s.s.p. of the system under full system restoration after its failure is represented by Equations 9.24 and 9.26. The Taylor expansion of these results up to the second order of the value $q = \max\{\alpha_1, \alpha_2\} \to 0$ allows us to prove the following theorem.

Theorem 9.7: Under components' rare failures, the s.s.p. of the system considered with full repair take the form

$$\pi_0 \approx \left[1+\rho_1(1+\alpha_2 b_3)+\rho_2(1+\alpha_1 b_3)\right]^{-1}$$

$$\pi_1 \approx \frac{\rho_1}{1+\rho_1(1+\alpha_2 b_3)+\rho_2(1+\alpha_1 b_3)}$$

$$\pi_2 \approx \frac{\rho_2}{1+\rho_1(1+\alpha_2 b_3)+\rho_2(1+\alpha_1 b_3)}$$

$$\pi_3 \approx \frac{(\rho_1\alpha_2+\rho_2\alpha_1)b_3}{1+\rho_1(1+\alpha_2 b_3)+\rho_2(1+\alpha_1 b_3)} \quad (9.34)$$

Proof. Expansion around zero of $\max\{\alpha_1, \alpha_2\}$ in the formula for π_0 in Equation 9.24 gives

$$\pi_0^{-1} = 1+\left(1-\tilde{b}_1(\alpha_2)\right)\left(\frac{\alpha_1}{\alpha_2}+\alpha_1 b_3\right)+\left(1-\tilde{b}_2(\alpha_1)\right)\left(\frac{\alpha_2}{\alpha_1}+\alpha_2 b_3\right)$$

$$\approx 1+\rho_1(1+\alpha_2 b_3)+\rho_2(1+\alpha_1 b_3)$$

Analogously, expansion around zero of $\max\{\alpha_1, \alpha_2\}$ in Equations (9.26) gives

$$\pi_1 = \frac{\alpha_1}{\alpha_2}\left(1-\tilde{b}_1(\alpha_2)\right)\pi_0 \approx \frac{\alpha_1}{\alpha_2}b_1\pi_0 = \rho_1\pi_0$$

$$\pi_2 = \frac{\alpha_2}{\alpha_1}\left(1-\tilde{b}_2(\alpha_1)\right)\pi_0 \approx \frac{\alpha_2}{\alpha_1}b_2\pi_0 = \rho_2\pi_0$$

$$\pi_3 = \left[\alpha_1\left(1-\tilde{b}_1(\alpha_2)\right) + \alpha_2\left(1-\tilde{b}_1(\alpha_2)\right)\right]b_3\pi_0 \approx (\alpha_2\rho_1 + \alpha_1\rho_2)b_3\pi_0$$

that coincide with (9.32). ∎

The results of the theorem show asymptotic insensitivity of the s.s.p. to the shapes of their components' repair distributions, but only to their mean values and the components' failure intensities.

9.6.3 Quasi-Stationary Probabilities

For q.s.p, taking into account that $\gamma < \min \alpha_i (i=1,2)$, with the help of the Taylor expansion of Equations (9.32) near the points $\alpha_i - \gamma$ when $\alpha = \max\{\alpha_i, i=1,2\} \to 0$, one may obtain the following theorem.

Theorem 9.8: Under components' rare failures, the q.s.p. of the system considered with full repair take the form

$$\pi_0 \approx (1+\rho_1+\rho_2)^{-1}$$

$$\pi_1 \approx \frac{\rho_1}{1+\rho_1+\rho_2}$$

$$\pi_2 \approx \frac{\rho_2}{1+\rho_1+\rho_2} \qquad (9.35)$$

Proof. For π_0 in (9.32) one has

$$\hat{\pi}_0 = \left[1 + \frac{\alpha_1}{\alpha_2-\gamma}\left(1-\tilde{b}_1(\alpha_2-\gamma)\right) + \frac{\alpha_2}{\alpha_1-\gamma}\left(1-\tilde{b}_2(\alpha_1-\gamma)\right)\right]^{-1}$$

$$\approx (1+\rho_1+\rho_2)^{-1}$$

while for π_i ($i=1,2$), taking into account that

$$\phi_i(-\gamma) = (\alpha_i-\gamma)\left(1-\tilde{b}_i(\alpha_{i^*}-\gamma)\right) \approx (\alpha_i-\gamma)b_i(\alpha_{i^*}-\gamma)$$

one may obtain

$$\hat{\pi}_i = \alpha_i \frac{\left(1-\tilde{b}_i(\alpha_{i^*}-\gamma)\right)(\alpha_i-\gamma)}{(\alpha_i-\gamma)(\alpha_{i^*}-\gamma) + \alpha_i\phi_i(-\gamma) + \alpha_{i^*}\phi_{i^*}(-\gamma)}$$

$$\approx \frac{\rho_i}{1+\rho_1+\rho_2}$$

which proves the theorem. ∎

9.7 Conclusion

The Markovization method was used for heterogeneous double redundant hot-standby renewable reliability system analysis. The time-dependent, stationary, and quasi-stationary probability distributions for the system were calculated. It was shown that under rare failures, the reliability characteristics are asymptotically insensitive to the shape of the elements' repair time distributions up to their two first moments.

Further investigations should be directed to the consideration of systems with more complex structure and systems with more complex failure processes, for example, the recurrent failure process or Marshall–Olkin type failure model.

9.8 Acknowledgments

This work was supported by the RUDN University Program 5-100 and was funded by the Russian Foundation for Basic Research under research projects No. 17-07-00142 and No. 17-01-00633.

The author thanks the anonymous reviewer, whose remarks promoted the improvement of the chapter.

References

1. B.A. Sevast'yanov (1957). An Ergodic theorem for Markov processes and its application to telephone systems with refusals. *Theory of Probability and its Applications*, Vol. 2, No. 1, pp. 104–112.
2. I.N. Kovalenko (1976). *Investigations on Analysis of Complex Systems Reliability*. Kiev: Naukova Dumka, 210 p. (In Russian).
3. V. Rykov (2013). Multidimensional alternative processes as reliability models. In: *Modern Probabilistic Methods for Analysis of Telecommunication Networks (BWWQT 2013), Proceedings*. Eds: A. Dudin, V. Klimenok, G. Tsarenkov, S. Dudin. Series: CCIS 356. Springer, New York, NY, pp. 147–157.
4. D. Koenig, V. Rykov, D. Schtoyn (1979). *Queueing Theory*. Gubkin University Press, Moscow, Russia p. 115 (in Russian). Moscow, Russia: Gubkin Russian State University.
5. B.V. Gnedenko (1964). On cold double redundant system. *Izv. AN SSSR. Texn. Cybern.* No. 4, pp. 3–12. (In Russian).
6. B.V. Gnedenko (1964). On cold double redundant system with restoration. *Izv. AN SSSR. Texn. Cybern.* No. 5, pp. 111–118. (In Russian).

7. A.D. Solov'ev (1970). On reservation with quick restoration. *Izv. AN SSSR. Texn. Cybern.* No. 1, pp. 56–71. (In Russian).
8. V. Rykov, Tran Ahn Ngia (2014). On sensitivity of systems reliability characteristics to the shape of their elements life and repair time distributions. *Vestnik PFUR. Ser. Mathematics. Informatics. Physics.* No. 3, pp. 65–77. (In Russian).
9. D. Efrosinin, V. Rykov (2014). Sensitivity analysis of reliability characteristics to the shape of the life and repair time distributions. In: Dudin A., Nazarov A., Yakupov R., Gortsev A. (eds) *Information Technologies and Mathematical Modeling. ITMM 2014. Communication in Computer and Information Science*, Vol. 487, pp. 101–112.
10. D. Efrosinin, V. Rykov and V. Vishnevskiy (2014). Sensitivity of reliability models to the shape of life and repair time distributions. In: *Proceedings of the 9th International Conference on Availability, Reliability and Security (ARES 2014)*, Fribourg, Switzerland, September 8–12, 2014, pp. 393–396. IEEE CPS, DOI 10.1109/ARES 2014.65.
11. D.V. Kozyrev (2011). Analysis of asymptotic behavior of reliability properties of redundant systems under the fast recovery. *Bulletin of People's Friendship University of Russia. Series "Mathematics. Information Sciences. Physics"* No. 3, pp. 49–57. (In Russian).
12. V. Rykov, D. Kozyrev (2015). On sensitivity of steady state probabilities of a cold redundant system to the shape of life and repair times distributions of its elements. In: *Proceedings of the Eighth International Workshop on Simulation*, Vienna, September 21–25, 2015. (under review)
13. V. Rykov, D. Kozyrev, E. Zaripova (2017). Modeling and simulation of reliability function of a homogeneous hot double redundant repairable system. In: *Proceedings 31st European Conference on Modeling and Simulation (ECMS 2017)*, Vienna, Austria. May 23–26, 2017. pp. 701–705.
14. V.V. Kalashnikov (1997). *Geometric Sums: Bounds for by Rare Events with Applications: Risk Analysis, Reliability, Queueing.* Dordrecht: Kluwer Academic Publishers, 1997. 256 p.
15. V. Rykov, D. Kozyrev (2017). Analysis of renewable reliability systems by Markovization method. *Analytical and Computational Methods in Probability Theory and its Applications (ACMPT-2017)*, Moscow, RUDN, 2017, pp. 727–734.
16. V. Rykov, D. Kozyrev (2017). Analysis of renewable reliability systems Markovization method. In: Rykov V., Singpurwalla N., Zubkov A. (eds) *Analytical and Computational Methods in Probability Theory* and *its Applications (ACMPT-2017)*, Lecture Notes in Computer Science, vol. 10684, Springer. Cham, pp. 210–220.
17. V. Rykov, D. Kozyrev. On reliability function of double redundant system with general repair time distributions. *Applied Stochastic Models in Business and Industry* (forthcoming).
18. W. Feller (1966). *An Introduction to Probability Theory and its Applications.* Vol. II. London: John Wiley & Sons.

10

Fuzzy Reliability of Systems Using Different Types of Level (λ, 1) Interval-Valued Fuzzy Numbers

Pawan Kumar and S. B. Singh

CONTENTS

10.1 Introduction 197
10.2 Definition 199
10.3 Arithmetic Operations on Level (λ, 1) Interval-Valued Fuzzy Numbers 202
10.4 Fuzzy Reliability of Systems Using Level (λ, 1) Interval-Valued Fuzzy Numbers 202
    10.4.1 Series System 202
    10.4.2 Parallel System 204
    10.4.3 Parallel–Series System 206
    10.4.4 Series–Parallel System 207
10.5 Discussion 212
10.6 Conclusion 212
References 212

10.1 Introduction

One of the important engineering tasks in design and development of any technical system is reliable engineering. Traditional reliability analysis has been found to be insufficient to handle uncertainty of data. For this reason, the fuzzy reliability concept has been introduced and formulated in the context of fuzzy measures. Onisawa and Kacprzyk (1995) have used fuzzy set theory in the evaluation of the reliability of a system. According to Zadeh (1965), in a fuzzy set, the degree of membership functions of an element in the universe has a single value: either 0 or 1e. Often, specialists are uncertain about the values of the membership of an element in a set. Hence, it is better to represent the values of the membership of an element in the set by intervals of possible real numbers instead of real numbers. To deal with these situations, the theory of interval-valued fuzzy set was proposed by

Gorzalczany (1987) and studied by Turksen (1996). Wang and Li (1998, 2001) provided extended operations of interval-valued fuzzy numbers and proposed the concept and properties of the similarity coefficient of the interval-valued fuzzy numbers. These types of fuzzy sets have been intensively investigated, not only in terms of theory, but also in terms of applications. Cai et al. (1991) presented the following two assumptions:

1. *Fuzzy-state assumption:* The meaning of system failure cannot be precisely defined in a reasonable way. At any time, the system may be in one of two fuzzy states: the fuzzy success state or the fuzzy failure state.
2. *Possibility assumption:* System behavior can be fully characterized in the context of possibility measures.

Cai et al. (1991) pointed out that fuzzy reliability can be physically interpreted as the probability that no capable performance decline occurs in a predefined time interval. Mon and Cheng (1994) presented a new method for fuzzy system reliability analysis for components with different membership functions. Mon and Chen (2003) used fuzzy numbers to find the fuzzy reliability of a series and parallel system. Aliev and Kara (2004) analyzed fuzzy system reliability using a time-dependent fuzzy set. Yao et al. (2008) used triangular fuzzy numbers and statistical data to find the fuzzy reliability of systems. Mahapatra et al. (2010) proposed fuzzy fault tree analysis using intuitionistic fuzzy numbers. Kumar et al. (2013) proposed a new approach for analysis of fuzzy system reliability using intuitionistic fuzzy numbers. Lee (2012) provided fuzzy parallel system reliability analysis based on level (λ, ρ) interval-valued fuzzy numbers. Wang et al. (2013) provided new operators on triangular intuitionistic fuzzy numbers and their applications in system fault analysis. Fuh (2014) used level $(\lambda, 1)$ interval-valued fuzzy numbers to evaluate the fuzzy reliability of systems. Chaube and Singh (2014a) analyzed the fuzzy reliability of different systems using conflicting bifuzzy sets. They assumed that the reliability of all components of a system follow such a conflicting bifuzzy set. Chaube and Singh (2014b) evaluated the fuzzy reliability of two-stage fuzzy weighted-k-out-of-n systems having common components.

In the abovementioned studies, the authors assumed that all components of a system follow the same type of fuzzy number or the level $(\lambda, 1)$ interval-valued fuzzy number. However, in practical problems, such types of situations probably do not occur. Hence, in contrast to earlier studies, we assumed that the reliability of all components of a system follow different types of level $(\lambda, 1)$ interval-valued fuzzy numbers (both triangular and trapezoidal). Using the proposed method, the fuzzy reliability of series, parallel, parallel–series, and series–parallel systems are evaluated. Numerical examples are also taken to illustrate the proposed study. The chapter is organized as follows: Section 10.2 defines the fuzzy set and interval-valued fuzzy numbers; Section 10.3 describes the arithmetical operations; Section 10.4 presents the evaluation of fuzzy reliability of systems; and Section 10.5 discusses the conclusions.

10.2 Definition

1. *Fuzzy set* (Zadeh 1965): If a set X be fixed, then a fuzzy set $\tilde{Z}$ is given by

$$\tilde{Z} = \{\langle x, \mu_{\tilde{Z}}(x) \rangle : x \in X\}$$

where $\mu_{\tilde{Z}}(x) \in [0,1]$ is the membership degree of the element $x \in X$.

2. *Interval-valued fuzzy number*: An interval-valued fuzzy set on R is given by

$$\tilde{A} = \{x, [\mu_{\tilde{A}^L}(x), \mu_{\tilde{A}^U}(x)]\}$$

where:

$$\mu_{\tilde{A}^L}(x) \leq \mu_{\tilde{A}^U}(x)$$

$$\mu_{\tilde{A}^L}(x), \mu_{\tilde{A}^U}(x) \in [0,1]$$

$$\mu_{\tilde{A}}(x) = [\mu_{\tilde{A}^L}(x), \mu_{\tilde{A}^U}(x)], \forall x \in R$$

or

$\tilde{A} = [\tilde{A}^L, \tilde{A}^U]$, where $\tilde{A}^L = \inf \tilde{A}$ and $\tilde{A}^U = \sup \tilde{A}$.

3. *Interval-valued triangular fuzzy number*:

Let $\tilde{A}^L = (a,b,c;\lambda)$ and $\tilde{A}^U = (e,b,h;\rho)$, $0 < \lambda \leq \rho \leq 1$ and $e < a < b < c < h$; then the interval-valued fuzzy set is expressed as a level (λ, ρ) interval-valued triangular fuzzy number and is also expressed as

$$\tilde{A} = \left[(a,b,c;\lambda),(e,b,h;\rho)\right].$$

The interval-valued fuzzy set $\tilde{A}$ indicates that, when the membership grade of X belongs to the interval $[\mu_{\tilde{A}^L}(x), \mu_{\tilde{A}^U}(x)]$, the smallest grade is $\mu_{\tilde{A}^L}(x)$ and the largest grade is $\mu_{\tilde{A}^U}(x)$.

Let $\tilde{A}^L = (a,b,c;\lambda)$, $a<b<c$ and $\tilde{A}^L = (p,b,r;\rho)$, $p<b<r$. Then,

$$\mu_{\tilde{A}^L}(x) = \begin{cases} \lambda \dfrac{(x-a)}{(b-a)}, & a \le x \le b \\ \lambda \dfrac{(c-x)}{(c-b)}, & b \le x \le c \\ 0, & \text{otherwise} \end{cases} \quad \mu_{\tilde{A}^U}(x) = \begin{cases} \rho \dfrac{(x-p)}{(b-p)}, & p \le x \le b \\ \rho \dfrac{(r-x)}{(r-b)}, & b \le x \le r \\ 0, & \text{otherwise} \end{cases}$$

the α-cuts of $\tilde{A}^L$ and $\tilde{A}^U$, ($0 \le \alpha \le 1$) are as follows:

For a triangular fuzzy number, if $0 \le \alpha < \lambda$, then the α-cut (shown in Figure 10.1) is given by

$$\tilde{A}_l^L(\alpha) = a + (b-a)\frac{\alpha}{\lambda}, \quad \tilde{A}_r^L(\alpha) = c - (c-b)\frac{\alpha}{\lambda}$$

$$A_l^U(\alpha) = p + (b-p)\frac{\alpha}{\lambda}, \quad A_r^U(\alpha) = r - (r-b)\frac{\alpha}{\lambda}$$

where $\tilde{A}_\alpha = \{[\tilde{A}_l^L(\alpha), \tilde{A}_r^L(\alpha)], [\tilde{A}_l^U(\alpha), \tilde{A}_r^U(\alpha)]\}$
and if $\lambda \le \alpha \le \rho$, then

$$A_l^U(\alpha) = p + (b-p)\frac{\alpha}{\lambda}, \quad A_r^U(\alpha) = r - (r-b)\frac{\alpha}{\lambda}$$

where $\tilde{A}_\alpha = [\tilde{A}_l^U(\alpha), \tilde{A}_r^U(\alpha)]$

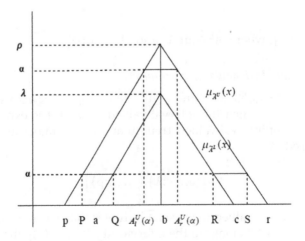

FIGURE 10.1
α-Cut of (λ, ρ) interval-valued triangular fuzzy numbers.

4. Interval-valued trapezoidal fuzzy number:

Let $\tilde{A}^L = (a,b,c,d;\lambda)$ and $\tilde{A}^U = (e,b,c,h;\rho)$, $0 < \lambda \leq \rho \leq 1$ and $e < a < b < c < d < h$. Then, the interval-valued fuzzy set is expressed as a level (λ, ρ) interval-valued trapezoidal fuzzy number and is also expressed as

$$\tilde{A} = \left[(a,b,c,d;\lambda),(e,b,c,h;\rho)\right]$$

The membership function of the smallest grade $\mu_{\tilde{A}^L}(x)$ and the largest grade $\mu_{\tilde{A}^U}(x)$ is expressed as

$$\mu_{\tilde{A}^L}(x) = \begin{cases} \lambda\frac{(x-a)}{(b-a)}, & a \leq x \leq b \\ \lambda, & b \leq x \leq c \\ \lambda\frac{(c-x)}{(d-c)}, & c \leq x \leq d \\ 0, & \text{otherwise} \end{cases}, \quad \mu_{\tilde{A}^U}(x) = \begin{cases} \rho\frac{(x-p)}{(b-p)}, & p \leq x \leq b \\ \rho, & b \leq x \leq c \\ \rho\frac{(r-x)}{(s-c)}, & c \leq x \leq s \\ 0, & \text{otherwise} \end{cases}$$

For a trapezoidal fuzzy number, if $0 \leq \alpha < \lambda$, then the α-cut is given by

$$A_l^L(\alpha) = a + (b-a)\frac{\alpha}{\lambda}, \quad A_r^L(\alpha) = c - (c-b)\frac{\alpha}{\lambda}$$

$$A_l^U(\alpha) = p + (b-p)\frac{\alpha}{\lambda}, \quad A_r^U(\alpha) = s - (s-b)\frac{\alpha}{\lambda}$$

where $\tilde{A}_\alpha = \{[\tilde{A}_l^L(\alpha), \tilde{A}_r^L(\alpha)], [\tilde{A}_l^U(\alpha), \tilde{A}_r^U(\alpha)]\}$
and if $\lambda \leq \alpha \leq \rho$, then

$$A_l^U(\alpha) = p + (b-p)\frac{\alpha}{\lambda}, \quad A_r^U(\alpha) = s - (s-b)\frac{\alpha}{\lambda}$$

where $\tilde{A}_\alpha = [\tilde{A}_l^U(\alpha), \tilde{A}_r^U(\alpha)]$.

The α-cut of a (λ, ρ) interval-valued trapezoidal fuzzy number is shown in Figure 10.2.

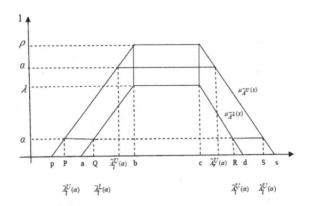

FIGURE 10.2
α-Cut of (λ, ρ) interval-valued trapezoidal fuzzy number.

10.3 Arithmetic Operations on Level (λ, 1) Interval-Valued Fuzzy Numbers

Let $\tilde{A}_\alpha = \{[\tilde{A}_l^L(\alpha), \tilde{A}_r^L(\alpha)], [\tilde{A}_l^U(\alpha), \tilde{A}_r^U(\alpha)]\}$ and $\tilde{B}_\alpha = \{[\tilde{B}_l^L(\alpha), \tilde{B}_r^L(\alpha)], \tilde{B}_\alpha = \{[\tilde{B}_l^L(\alpha), \tilde{B}_r^L(\alpha)], [\tilde{B}_l^U(\alpha), \tilde{B}_r^U(\alpha)]\}$; then,

1. Addition:

$$\tilde{A}_\alpha + \tilde{B}_\alpha = \left\{[\tilde{A}_l^L(\alpha) + \tilde{B}_l^L(\alpha), \tilde{A}_r^L(\alpha) + \tilde{B}_r^L(\alpha)], [\tilde{A}_l^U(\alpha) + \tilde{B}_l^U(\alpha), \tilde{A}_r^U(\alpha) + \tilde{B}_r^U(\alpha)]\right\}$$

2. Subtraction:

$$\tilde{A}_\alpha - \tilde{B}_\alpha = \left\{[\tilde{A}_l^L(\alpha) - \tilde{B}_l^L(\alpha), \tilde{A}_r^L(\alpha) - \tilde{B}_r^L(\alpha)], [\tilde{A}_l^U(\alpha) - \tilde{B}_l^U(\alpha), \tilde{A}_r^U(\alpha) - \tilde{B}_r^U(\alpha)]\right\}$$

3. Complement:

$$1 - \tilde{A}_\alpha = \left\{[1 - \tilde{A}_r^U(\alpha), 1 - \tilde{A}_l^U(\alpha)], [1 - \tilde{A}_r^L(\alpha), 1 - \tilde{A}_l^L(\alpha)]\right\}$$

10.4 Fuzzy Reliability of Systems Using Level (λ, 1) Interval-Valued Fuzzy Numbers

10.4.1 Series System

Consider a series system consisting of n components as depicted in Figure 10.3. The fuzzy reliability $\tilde{R}_S$ of the series system having the reliability of each

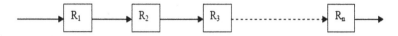

FIGURE 10.3
Series system.

component as different types of level $(\lambda, 1)$ interval-valued fuzzy number can be evaluated by using the following expression.

Fuzzy reliability:

$$\tilde{R}_s = \underset{j=1}{\overset{n}{\otimes}} \tilde{R}_j = \tilde{R}_1 \otimes \tilde{R}_2 \otimes \tilde{R}_3 \ldots \ldots \tilde{R}_n$$

Numerical example: Consider a radio set consisting of three independent major components: a power supply, a receiver, and an amplifier all work normally to operate the radio set successfully. Let the reliability of each component be taken as different types of level $(\lambda, 1)$ interval-valued fuzzy numbers $\tilde{R}_1, \tilde{R}_2,$ and $\tilde{R}_3$. In each case, three different possibilities of level $(\lambda, 1)$ interval-valued fuzzy number have been considered corresponding to three different components, as shown in Table 10.1.

The α-cut of fuzzy reliability of the series system when $0 \le \alpha < \lambda = 0.9$ and $\lambda \le \alpha \le \rho = 1$ are given in Tables 10.2 and 10.3, respectively.

TABLE 10.1

Level $(\lambda, 1)$ Interval-Valued Fuzzy Reliability of Components in Series System

	Reliability of ith Component in Series System	Type of Fuzzy Number
Case I	$\tilde{R}_1 = [(0.02, 0.05, 0.08; 0.9), (0.01, 0.05, 0.09; 1)]$	Level $(\lambda, 1)$ interval-valued triangular fuzzy number
	$\tilde{R}_2 = [(0.03, 0.05, 0.07, 0.09; 0.9), (0.02, 0.05, 0.07, 0.11; 1)]$	Level $(\lambda, 1)$ interval-valued trapezoidal fuzzy number
	$\tilde{R}_3 = [(0.02, 0.04, 0.07; 0.9), (0.01, 0.04, 0.08; 1)]$	Level $(\lambda, 1)$ interval-valued triangular fuzzy number
Case II	$\tilde{R}_1 = [(0.02, 0.05, 0.08, 0.11; 0.9), (0.01, 0.05, 0.08, 0.13; 1)]$	Level $(\lambda, 1)$ interval-valued trapezoidal fuzzy number
	$\tilde{R}_2 = [(0.03, 0.05, 0.07; 0.9), (0.02, 0.05, 0.08; 1)]$	Level $(\lambda, 1)$ interval-valued triangular fuzzy number
	$\tilde{R}_3 = [(0.02, 0.04, 0.06, 0.08; 0.9), (0.01, 0.04, 0.06, 0.09; 1)]$	Level $(\lambda, 1)$ interval-valued trapezoidal fuzzy number
Case III	$\tilde{R}_1 = [(0.02, 0.05, 0.08; 0.9), (0.01, 0.05, 0.09; 1)]$	Level $(\lambda, 1)$ interval-valued triangular fuzzy number
	$\tilde{R}_2 = [(0.03, 0.05, 0.07; 0.9), (0.02, 0.05, 0.08; 1)]$	Level $(\lambda, 1)$ interval-valued triangular fuzzy number
	$\tilde{R}_3 = [(0.02, 0.04, 0.07, 0.09; 0.9), (0.01, 0.04, 0.07, 0.1; 1)]$	Level $(\lambda, 1)$ interval-valued triangular fuzzy number

TABLE 10.2

α-Cut of Series System Reliability When $0 \le \alpha < \lambda = 0.9$

	α-Cut	$\tilde{R}_{s\alpha}$
Case I	0.0	{[0.000012, 0.000504], [0.000002, 0.000792]}
	0.1	{[0.0000165, 0.000449], [0.00000419, 0.000693]}
	0.2	{[0.0000223, 0.000398], [0.00000748, 0.0006022]}
	0.3	{[0.0000291, 0.000351], [0.00001212, 0.0005198]}
	0.4	{[0.0000371, 0.000308], [0.0000183, 0.0004452]}
	0.5	{[0.0000434, 0.000268], [0.0000263, 0.000378]}
	0.6	{[0.0000571, 0.000232], [0.00003617, 0.0003179]}
	0.7	{[0.0000693, 0.000199], [0.0000483, 0.0002644]}
	0.8	{[0.00008305, 0.0001692], [0.0000628, 0.000217]}
Case II	0.0	{[0.000012, 0.000616], [0.000002, 0.000936]}
	0.1	{[0.00001665, 0.0005628], [0.00000446, 0.0008279]}
	0.2	{[0.0000223, 0.0005128], [0.000008301, 0.0007285]}
	0.3	{[0.00002911, 0.0004658], [0.000013804, 0.0006373]}
	0.4	{[0.0000371, 0.0004218], [0.000021258, 0.00055406]}
	0.5	{[0.0000464, 0.0003806], [0.00003095, 0.00047839]}
	0.6	{[0.0000571, 0.00034224], [0.000043172, 0.00040993]}
	0.7	{[0.0000693, 0.0003065], [0.00005821, 0.0003483]}
	0.8	{[0.00008304, 0.00027335], [0.00007634, 0.0002932]}
Case III	0.0	{[0.000012, 0.000504], [0.000002, 0.00072]}
	0.1	{[0.0000167, 0.0004566], [0.00000434, 0.0006378]}
	0.2	{[0.00002304, 0.00041217], [0.00000795, 0.00056215]}
	0.3	{[0.00003126, 0.00037066], [0.00001305, 0.00049265]}
	0.4	{[0.00004188, 0.00033196], [0.00002003, 0.00042906]}
	0.5	{[0.00005537, 0.00029597], [0.00002902, 0.00037115]}
	0.6	{[0.00007217, 0.0002626], [0.00004033, 0.0003186]}
	0.7	{[0.00009275, 0.00023176], [0.00005421, 0.0002713]}
	0.8	{[0.00011755, 0.00020335], [0.000070936, 0.0002288]}

TABLE 10.3

α-Cut of Series System Reliability When $\lambda \le \alpha \le \rho = 1$

α-Cut	Case I $\tilde{R}_{s\alpha}$	Case II $\tilde{R}_{s\alpha}$	Case III $\tilde{R}_{s\alpha}$
0.8	[0.00008083, 0.00011885]	[0.0000628, 0.00033264]	[0.00006283, 0.00024685]
0.9	[0.00009799, 0.00006177]	[0.00007999, 0.0002838]	[0.00007999, 0.00020893]
1.0	[0.000118, 0.000012]	[0.0001, 0.00024]	[0.0001, 0.000175]

10.4.2 Parallel System

Let us consider a parallel system consisting of n components, as shown in Figure 10.4. The fuzzy reliability $\tilde{R}_p$ of the parallel system can be evaluated by using the expression as follows:

$$\tilde{R}_p = 1 - \bigotimes_{j=1}^{n}\left(1 - \tilde{R}_j\right)$$

$$= 1 - \left(1 - \tilde{R}_1\right) \otimes \left(1 - \tilde{R}_2\right) \otimes \left(1 - \tilde{R}_3\right) \ldots \left(1 - \tilde{R}_n\right)$$

Fuzzy Reliability of Systems

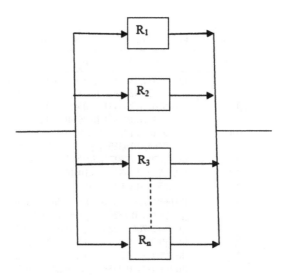

FIGURE 10.4
Parallel system.

Numerical example: Consider a space vehicle with three independent engines. At least one of these must function to operate the space vehicle to achieve orbit. Let the reliability of the engines be in the form of different types of level $(\lambda, 1)$ interval-valued fuzzy numbers, as given in three different cases of Table 10.4. In each case, three different possibilities of level

TABLE 10.4

Level $(\lambda, 1)$ Interval-Valued Fuzzy Reliability of Components in Parallel System

	Reliability of ith Component in Parallel System	Type of Fuzzy Number
Case I	$\tilde{R}_1 = [(0.02, 0.05, 0.08; 0.9), (0.01, 0.05, 0.09; 1)]$	Level $(\lambda, 1)$ interval-valued triangular fuzzy number
	$\tilde{R}_2 = [(0.03, 0.05, 0.07, 0.09; 0.9), (0.02, 0.05, 0.07, 0.11; 1)]$	Level $(\lambda, 1)$ interval-valued trapezoidal fuzzy number
	$\tilde{R}_3 = [(0.02, 0.04, 0.07; 0.9), (0.01, 0.04, 0.08; 1)]$	Level $(\lambda, 1)$ interval-valued triangular fuzzy number
Case II	$\tilde{R}_1 = [(0.02, 0.05, 0.08, 0.11; 0.9), (0.01, 0.05, 0.08, 0.13; 1)]$	Level $(\lambda, 1)$ interval-valued trapezoidal fuzzy number
	$\tilde{R}_2 = [(0.03, 0.05, 0.07; 0.9), (0.02, 0.05, 0.08; 1)]$	Level $(\lambda, 1)$ interval-valued triangular fuzzy number
	$\tilde{R}_3 = [(0.02, 0.04, 0.06, 0.08; 0.9), (0.01, 0.04, 0.06, 0.09; 1)]$	Level $(\lambda, 1)$ interval-valued trapezoidal fuzzy number
Case III	$\tilde{R}_1 = [(0.02, 0.05, 0.08; 0.9), (0.01, 0.05, 0.09; 1)]$	Level $(\lambda, 1)$ interval-valued triangular fuzzy number
	$\tilde{R}_2 = [(0.03, 0.05, 0.07; 0.9), (0.02, 0.05, 0.08; 1)]$	Level $(\lambda, 1)$ interval-valued triangular fuzzy number
	$\tilde{R}_3 = [(0.02, 0.04, 0.07, 0.09; 0.9), (0.01, 0.04, 0.07, 0.1; 1)]$	Level $(\lambda, 1)$ interval-valued trapezoidal fuzzy number

TABLE 10.5

α-Cut of Parallel System Reliability When $0 \le \alpha < \lambda = 0.9$

	α-Cut	$\tilde{R}_{p\alpha}$
Case I	0.0	{[0.068412, 0.221404], [0.039502, 0.254892]}
	0.1	{[0.07573, 0.21394], [0.049201, 0.24499]}
	0.2	{[0.08302, 0.20644], [0.058836, 0.23499]}
	0.3	{[0.09027, 0.19889], [0.068406, 0.22493]}
	0.4	{[0.09748, 0.191296], [0.077914, 0.21477]}
	0.5	{[0.10465, 0.183656], [0.087356, 0.20454]}
	0.6	{[0.11179, 0.175972], [0.0967356, 0.194213]}
	0.7	{[0.11889, 0.16824], [0.106051, 0.183805]}
	0.8	{[0.12595, 0.160467], [0.115304, 0.173313]}
Case II	0.0	{[0.068412, 0.238516], [0.039502, 0.263719]}
	0.1	{[0.076449, 0.232053], [0.0501674, 0.2537389]}
	0.2	{[0.08423706, 0.225554], [0.0607536, 0.243673]}
	0.3	{[0.0917723, 0.2190188], [0.0712606, 0.2335197]}
	0.4	{[0.0990347, 0.2124484], [0.0816879, 0.2232801]}
	0.5	{[0.1060834, 0.205842], [0.0920354, 0.2129534]}
	0.6	{[0.1128574, 0.1991998], [0.1023026, 0.2025389]}
	0.7	{[0.1193756, 0.1925214], [0.1124893, 0.1920368]}
	0.8	{[0.1256373, 0.1858069], [0.1225952, 0.1814463]}
Case III	0.0	{[0.068412, 0.221404], [0.039502, 0.24652]}
	0.1	{[0.0755286, 0.214869], [0.048066, 0.237375]}
	0.2	{[0.082196, 0.208298], [0.056572, 0.2281569]}
	0.3	{[0.088415, 0.201692], [0.065022, 0.2188657]}
	0.4	{[0.094185, 0.1950502], [0.073415, 0.209501]}
	0.5	{[0.099506, 0.188372], [0.081752, 0.2000623]}
	0.6	{[0.1043795, 0.1816576], [0.090032, 0.1905496]}
	0.7	{[0.108804, 0.174907], [0.098256, 0.1809625]}
	0.8	{[0.1127808, 0.1681203], [0.106425, 0.17130068]}

TABLE 10.6

α-Cut of Parallel System Reliability When $\lambda \le \alpha \le \rho = 1$

α-Cut	Case I $\tilde{R}_{p\alpha}$	Case II $\tilde{R}_{p\alpha}$	Case III $\tilde{R}_{p\alpha}$
0.8	[0.117166, 0.173313]	[0.1152908, 0.1783348]	[0.1253972, 0.1783348]
0.9	[0.12685, 0.162734]	[0.124476, 0.1695359]	[0.1344252, 0.1695359]
1.0	[0.136531, 0.15207]	[0.1336, 0.160675]	[0.143278, 0.160675]

$(\lambda, 1)$ interval-valued fuzzy numbers have been considered, corresponding to three different components.

The α-cut of the fuzzy reliability of the parallel system when $0 \le \alpha < \lambda = 0.9$ and $\lambda \le \alpha \le \rho = 1$ are given in Tables 10.5 and 10.6, respectively.

10.4.3 Parallel–Series System

Consider a parallel–series system consisting of m connections connected in parallel where each connection contains n components, as depicted in Figure 10.5. The fuzzy reliability $\tilde{R}_{ps}$ in this case is given by

Fuzzy Reliability of Systems

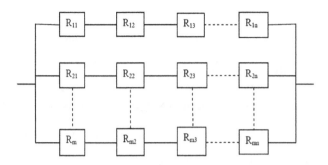

FIGURE 10.5
Parallel–series system.

$$\tilde{R}_{ps} = 1 - \overset{m}{\underset{i=1}{\otimes}} \left(1 - \left(\overset{n}{\underset{i=1}{\otimes}} \tilde{R}_{ki} \right) \right)$$

where $\tilde{R}_{ki}$ represents the reliability of the ith component in the kth network.

Numerical example: Consider a communication system that receives an input signal and transmits an output signal. Let there be two receivers and two transmitters in the system connected in a parallel–series system. For successful communication, at least one receiver and one transmitter connected in a series configuration must work properly. The reliability of receivers and transmitters are in the form of different types of level $(\lambda, 1)$ interval-valued fuzzy numbers, as given in three different cases of Table 10.7. In each case, three different possibilities of level $(\lambda, 1)$ interval-valued fuzzy numbers have been considered, corresponding to three different components.

The fuzzy reliability $\tilde{R}_{PS}$ of such a parallel–series network is given by

$$\tilde{R}_{PS} = 1 - \overset{m}{\underset{i=1}{\otimes}} \left(1 - \left(\overset{n}{\underset{k=1}{\otimes}} \tilde{R}_{ki} \right) \right)$$

$$= 1 - \left\{ \left[1 - \tilde{R}_{11}\tilde{R}_{21} \right] \otimes \left[1 - \tilde{R}_{12}\tilde{R}_{22} \right] \right\}$$

The α-cut of fuzzy reliability of the parallel–series system when $0 \le \alpha < \lambda = 0.9$ and $\lambda \le \alpha \le \rho = 1$ are given in Tables 10.8 and 10.9, respectively.

10.4.4 Series–Parallel System

Consider a series–parallel system that consists of n subsystems connected in series, where each subsystem contains m components as shown in Figure 10.6; then, the fuzzy reliability $\tilde{R}_{SP}$ of the system can be calculated as

$$\tilde{R}_{SP} = \overset{n}{\underset{k=1}{\otimes}} \left(1 - \overset{m}{\underset{i=1}{\otimes}} \left(1 - \tilde{R}_{ik} \right) \right)$$

where $\tilde{R}_{ik}$ represents the reliability of the kth component at the ith stage.

TABLE 10.7

Level (λ, 1) Interval-Valued Fuzzy Reliability of Components in Parallel–Series System

	Reliability of (i,j)th Component in Parallel–Series System	Type of Fuzzy Number
Case I	$\tilde{R}_{11} = [(0.02, 0.05, 0.08; 0.9), (0.01, 0.05, 0.09; 1)]$	Level (λ, 1) interval-valued triangular fuzzy number
	$\tilde{R}_{12} = [(0.03, 0.05, 0.07, 0.09; 0.9), (0.02, 0.05, 0.07, 0.11; 1)]$	Level (λ, 1) interval-valued trapezoidal fuzzy number
	$\tilde{R}_{21} = [(0.02, 0.04, 0.07; 0.9), (0.01, 0.04, 0.08; 1)]$	Level (λ, 1) interval-valued triangular fuzzy number
	$\tilde{R}_{22} = [(0.03, 0.08, 0.13, 0.18; 0.9), (0.05, 0.08, 0.13, 0.23; 1)]$	Level (λ, 1) interval-valued trapezoidal fuzzy number
Case II	$\tilde{R}_{11} = [(0.02, 0.05, 0.08; 0.9), (0.01, 0.05, 0.09; 1)]$	Level (λ, 1) interval-valued triangular fuzzy number
	$\tilde{R}_{21} = [(0.03, 0.05, 0.07; 0.9), (0.02, 0.05, 0.08; 1)]$	Level (λ, 1) interval-valued triangular fuzzy number
	$\tilde{R}_{12} = [(0.02, 0.04, 0.07, 0.09; 0.9), (0.01, 0.04, 0.07, 0.11; 1)]$	Level (λ, 1) interval-valued trapezoidal fuzzy number
	$\tilde{R}_{22} = [(0.03, 0.08, 0.13, 0.18; 0.9), (0.05, 0.08, 0.13, 0.23; 1)]$	Level (λ, 1) interval-valued trapezoidal fuzzy number
Case III	$\tilde{R}_{1} = [(0.02, 0.05, 0.08; 0.9), (0.01, 0.05, 0.09; 1)]$	Level (λ, 1) interval-valued triangular fuzzy number
	$\tilde{R}_{2} = [(0.03, 0.05, 0.07; 0.9), (0.02, 0.05, 0.08; 1)]$	Level (λ, 1) interval-valued triangular fuzzy number
	$\tilde{R}_{3} = [(0.02, 0.04, 0.07; 0.9), (0.01, 0.04, 0.08; 1)]$	Level (λ, 1) interval-valued triangular fuzzy number
	$\tilde{R}_{22} = [(0.03, 0.08, 0.13, 0.18; 0.9), (0.05, 0.08, 0.13, 0.23; 1)]$	Level (λ, 1) interval-valued trapezoidal fuzzy number

Numerical example: Suppose there is a communication system that receives an input signal and transmits an output signal. For this, there are two transmitters in the system connected in a series–parallel system. For a successful communication, at least one receiver and one transmitter must work properly. Let the reliability of receivers and transmitters be in the form of different types of level (λ, 1) interval-valued fuzzy numbers, as given in three different cases in Table 10.10. In each case, three different possibilities of level (λ, 1) interval-valued fuzzy numbers have been considered, corresponding to three different components.

The fuzzy reliability $\tilde{R}_{SP}$ of this series–parallel network is given by

$$\tilde{R}_{SP} = \bigotimes_{k=1}^{n} \left(1 - \bigotimes_{i=1}^{m} \left(1 - \tilde{R}_{ik} \right) \right)$$

$$= \left[1 - \left(1 - \tilde{R}_{11} \right) \otimes \left(1 - \tilde{R}_{21} \right) \right] \otimes \left[1 - \left(1 - \tilde{R}_{12} \right) \otimes \left(1 - \tilde{R}_{22} \right) \right]$$

TABLE 10.8

α-Cut of Reliability of Parallel–Series System When $0 \leq \alpha < \lambda = 0.9$

	α-Cut	$\tilde{R}_{ps\alpha}$
Case I	0.0	{[0.0012, 0.01971], [0.0007, 0.02812]}
	0.1	{[0.001538, 0.0183], [0.00104, 0.026855]}
	0.2	{[0.001915, 0.01699], [0.001428, 0.02566]}
	0.3	{[0.00233, 0.01571], [0.001872, 0.024536]}
	0.4	{[0.002784, 0.014471], [0.002371, 0.023483]}
	0.5	{[0.003276, 0.01328], [0.002923, 0.0225]}
	0.6	{[0.003807, 0.012142], [0.00353, 0.021588]}
	0.7	{[0.004375, 0.011051], [0.004188, 0.020747]}
	0.8	{[0.004982, 0.010008], [0.004902, 0.0109977]}
Case II	0.0	{[0.0012, 0.021709], [0.0007, 0.032318]}
	0.1	{[0.00153813, 0.0204414], [0.0010443, 0.0315256]}
	0.2	{[0.001914875, 0.0192116], [0.0014393, 0.0308554]}
	0.3	{[0.00233022, 0.0180195], [0.00188508, 0.0303073]}
	0.4	{[0.002784145, 0.01686523], [0.00238157, 0.0298819]}
	0.5	{[0.003276618, 0.0157488], [0.0029287, 0.0295796]}
	0.6	{[0.00380762, 0.0146705], [0.0035265, 0.0294006]}
	0.7	{[0.004377107, 0.0136301], [0.00417532, 0.029345]}
	0.8	{[0.00498506, 0.0126276], [0.0048744, 0.0294135]}
Case III	0.0	{[0.0012, 0.01813], [0.0007, 0.0255]}
	0.1	{[0.001538, 0.016779], [0.0014427, 0.0251619]}
	0.2	{[0.001915, 0.015479], [0.0014393, 0.024785]}
	0.3	{[0.00233, 0.014229], [0.0018849, 0.0243699]}
	0.4	{[0.002784, 0.0130305], [0.0023814, 0.023916]}
	0.5	{[0.003276, 0.01188323], [0.0029284, 0.023424]}
	0.6	{[0.003807, 0.0107879], [0.0035259, 0.0228929]}
	0.7	{[0.004376, 0.0097449], [0.004174, 0.0223235]}
	0.8	{[0.004984, 0.0087549], [0.0048727, 0.0217156]}

TABLE 10.9

α-Cut of Reliability of Parallel–Series System When $\lambda \leq \alpha \leq \rho = 1$

α-Cut	Case I $\tilde{R}_{ps\alpha}$	Case II $\tilde{R}_{ps\alpha}$	Case III $\tilde{R}_{ps\alpha}$
0.8	[0.006697, 0.013288]	[0.0043595, 0.0149116]	[0.004362, 0.010424]
0.9	[0.007463, 0.011705]	[0.0050049, 0.0131939]	[0.0050084, 0.0090033]
1.0	[0.008283, 0.010182]	[0.0056921, 0.0115787]	[0.0056971, 0.0076854]

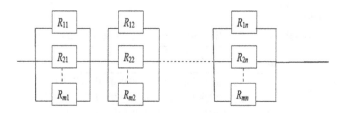

FIGURE 10.6
Series–parallel system.

TABLE 10.10

Level (λ, 1) Interval-Valued Fuzzy Reliability of Components in Series–Parallel System

	Reliability of (i,j)th Component in Series–Parallel System	Types of Fuzzy Number
Case I	$\tilde{R}_{11} = [(0.02, 0.05, 0.08; 0.9), (0.01, 0.05, 0.09; 1)]$	Level (λ, 1) interval-valued triangular fuzzy number
	$\tilde{R}_{12} = [(0.03, 0.05, 0.07, 0.09; 0.9), (0.02, 0.05, 0.07, 0.11; 1)]$	Level (λ, 1) interval-valued trapezoidal fuzzy number
	$\tilde{R}_{21} = [(0.02, 0.04, 0.07; 0.9), (0.01, 0.04, 0.08; 1)]$	Level (λ, 1) interval-valued triangular fuzzy number
	$\tilde{R}_{22} = [(0.03, 0.08, 0.13, 0.18; 0.9), (0.05, 0.08, 0.13, 0.23; 1)]$	Level (λ, 1) interval-valued trapezoidal fuzzy number
Case II	$\tilde{R}_{11} = [(0.02, 0.05, 0.08; 0.9), (0.01, 0.05, 0.09; 1)]$	Level (λ, 1) interval-valued triangular fuzzy number
	$\tilde{R}_{21} = [(0.03, 0.05, 0.07; 0.9), (0.02, 0.05, 0.08; 1)]$	Level (λ, 1) interval-valued triangular fuzzy number
	$\tilde{R}_{12} = [(0.02, 0.04, 0.07, 0.09; 0.9), (0.01, 0.04, 0.07, 0.11; 1)]$	Level (λ, 1) interval-valued trapezoidal fuzzy number
	$\tilde{R}_{22} = [(0.03, 0.08, 0.13, 0.18; 0.9), (0.05, 0.08, 0.13, 0.23; 1)]$	Level (λ, 1) interval-valued trapezoidal fuzzy number
Case III	$\tilde{R}_1 = [(0.02, 0.05, 0.08; 0.9), (0.01, 0.05, 0.09; 1)]$	Level (λ, 1) interval-valued triangular fuzzy number
	$\tilde{R}_2 = [(0.03, 0.05, 0.07; 0.9), (0.02, 0.05, 0.08; 1)]$	Level (λ, 1) interval-valued triangular fuzzy number
	$\tilde{R}_3 = [(0.02, 0.04, 0.07; 0.9), (0.01, 0.04, 0.08; 1)]$	Level (λ, 1) interval-valued triangular fuzzy number
	$\tilde{R}_{22} = [(0.03, 0.08, 0.13, 0.18; 0.9), (0.05, 0.08, 0.13, 0.23; 1)]$	Level (λ, 1) interval-valued trapezoidal fuzzy number

The α-cuts of fuzzy reliability of this series–parallel system when $0 \le \alpha < \lambda = 0.9$ and $\lambda \le \alpha \le \rho = 1$ are given in Tables 10.11 and 10.12, respectively.

TABLE 10.11

α-Cut of Reliability of Series–Parallel System When $0 \leq \alpha < \lambda = 0.9$

	α-Cut	$\tilde{R}_{sp\alpha}$
Case I	0.0	{[0.002340, 0.036648], [0.001373, 0.051233]}
	0.1	{[0.002994, 0.034153], [0.002058, 0.047755]}
	0.2	{[0.037240, 0.031733], [0.002846, 0.044372]}
	0.3	{[0.004533, 0.029388], [0.003738, 0.041083]}
	0.4	{[0.005418, 0.027122], [0.004730, 0.037891]}
	0.5	{[0.006378, 0.024932], [0.005825, 0.034795]}
	0.6	{[0.007415, 0.022820], [0.007018, 0.031798]}
	0.7	{[0.008526, 0.020789], [0.008309, 0.028899]}
	0.8	{[0.009712, 0.018838], [0.009697, 0.026102]}
Case II	0.0	{[0.00244, 0.03665], [0.001773, 0.051233]}
	0.1	{[0.003116, 0.034389], [0.0024619, 0.04695]}
	0.2	{[0.003869, 0.032161], [0.0032443, 0.042831]}
	0.3	{[0.0046999, 0.029968], [0.0041193, 0.0388797]}
	0.4	{[0.0056073, 0.0278147], [0.0050859, 0.0350988]}
	0.5	{[0.0065905, 0.0257036], [0.0061429, 0.0314918]}
	0.6	{[0.0076489, 0.0236385], [0.0072894, 0.0280618]}
	0.7	{[0.0087817, 0.0216228], [0.0085243, 0.0248124]}
	0.8	{[0.0099883, 0.0196599], [0.0098465, 0.0217467]}
Case III	0.0	{[0.0024404, 0.03428], [0.001773, 0.04747]}
	0.1	{[0.0030611, 0.032067], [0.0024618, 0.0432987]}
	0.2	{[0.0037598, 0.029919], [0.0032443, 0.0392968]}
	0.3	{[0.0045358, 0.027838], [0.00411928, 0.03546774]}
	0.4	{[0.0053884, 0.0258242], [0.0050858, 0.03181465]}
	0.5	{[0.0063168, 0.0238778], [0.0061428, 0.0283409]}
	0.6	{[0.0073203, 0.02200004], [0.0072893, 0.0250497]}
	0.7	{[0.00839831, 0.0201919], [0.008524113, 0.0219444]}
	0.8	{[0.00955003, 0.0184544], [0.00984635, 0.01902835]}

TABLE 10.12

α-Cut of Reliability of Series–Parallel System When $\lambda \leq \alpha \leq \rho = 1$

α-Cut	Case I $\tilde{R}_{sp\alpha}$	Case II $\tilde{R}_{sp\alpha}$	Case III $\tilde{R}_{sp\alpha}$
0.8	[0.0069975, 0.026366]	[0.008875, 0.0188513]	[0.00884964, 0.0261873]
0.9	[0.0111815, 0.023721]	[0.010095, 0.0152309]	[0.01009438, 0.0238905]
1.0	[0.0127602, 0.021184]	[0.011386, 0.0117929]	[0.01138482, 0.0216782]

10.5 Discussion

In this study, we have used level (λ, 1) interval-valued fuzzy numbers as the reliability corresponding to the different components of series, parallel, parallel–series, and series–parallel systems. In each case, three different possibilities have been considered. In this chapter, we observed the following:

1. *Series system:* The reliability of the series system is at its maximum in case II when $0 \leq \alpha < \lambda = 0.9$ and $\lambda \leq \alpha \leq \rho = 1$.
2. *Parallel system:* The reliability of the parallel system is at its maximum in case II when $0 \leq \alpha < \lambda = 0.9$ and the system is more reliable in case III when $\lambda \leq \alpha \leq \rho = 1$.
3. *Parallel–series system:* The reliability of the parallel–series system is at its maximum in case II when $0 \leq \alpha < \lambda = 0.9$ and $\lambda \leq \alpha \leq \rho = 1$.
4. *Series–parallel system:* The reliability of the series–parallel system is at its maximum in case I when $0 \leq \alpha < \lambda = 0.9$ and the system is more reliable in case III when $\lambda \leq \alpha \leq \rho = 1$.

10.6 Conclusion

In this study, we have defined level (λ, 1) interval-valued fuzzy numbers (both triangular and trapezoidal). Some arithmetical operations of the proposed level (λ, 1) interval-valued fuzzy numbers are also evaluated, based on the α-cut method. We have evaluated the fuzzy reliability of systems using the α-cut method and assumed that the reliability of each component of the systems follow different types of level (λ, 1) interval-valued fuzzy numbers (both triangular and trapezoidal) with three different possibilities, in contrast to the same type used in earlier studies. We observed that the fuzzy reliabilities of these systems are obtained in the form of trapezoidal level (λ, 1) interval-valued fuzzy numbers. The reliability in cases II and III of the parallel system is greater than in other cases when $0 \leq \alpha < \lambda = 0.9$ and $\lambda \leq \alpha \leq \rho = 1$, respectively. When $0 \leq \alpha < \lambda$, then fuzzy reliability is in the form of an interval-valued fuzzy number, and when $\lambda \leq \alpha \leq \rho = 1$, fuzzy reliabilities are obtained in the form of fuzzy numbers only.

References

Aliev M. and Kara Z., Fuzzy system reliability analysis using time dependent fuzzy set, *Control and Cybernetics*, (2004), 33:653–662.

Cai K. Y., Wen C. Y., and Zhang M. L., Fuzzy variables as a basis for a theory of fuzzy reliability in the possibility context, *Fuzzy Sets and Systems*, (1991), 42:142–145.

Chaube S. and Singh S. B., Fuzzy reliability evaluation using conflicting bifuzzy approach, ICDMIC-2014 proceedings to appear in *IEEE Xplore* (2014a), DOI:10.1109/ICDMIC.2014.6954241.

Chaube S. and Singh S. B., Fuzzy reliability evaluation of two-stage fuzzy weighted-k-out-of-n systems having common components, *Engineering and Automation Problems*, (2014b), 4:112–116.

Chen, S. M., Analyzing fuzzy system reliability using vague set theory, *International Journal of Applied Science and Engineering*, (2003), 1:82–88.

Fuh C. N., Jea R., and Su J. S., Fuzzy system reliability analysis based on level (λ, 1) interval-valued fuzzy numbers, *Information Sciences*, (2014), 272:185–197.

Gorzalczany M. B., A method of inference in approximate reasoning based on interval-valued fuzzy sets, *Fuzzy Sets and Systems*, (1987), 21:1–17.

Kumar M., Yadav. S. P., and Kumar S., Fuzzy system reliability evaluation using time-dependent intuitionistic fuzzy set, *International Journal of System Science*, (2013), 44:50–66.

Lee H. M., Fuh C. N., and Su J. S., Fuzzy parallel system reliability analysis based on level (λ, ρ) interval-valued fuzzy numbers, *International Journal of Innovative Computing, Information and Control*, (2012), 8:5703–5713.

Mahapatra G. S., Intuitionistic fuzzy fault tree analysis using intuitionistic fuzzy numbers, *International Mathematical Forum*, (2010), 21:1015–1024.

Mon D. L. and Cheng C. H., Fuzzy system reliability analysis for components with different membership functions, *Fuzzy Sets and Systems*, (1994), 64:145–157.

Onisawa T. and Kacprzyk J., *Reliability and Safety Analysis Under Fuzziness*, Physica-Verlag, Heidelberg, (1995).

Singer D., A fuzzy set approach to fault tree and reliability analysis, *Fuzzy Sets and Systems*, (1990), 34:145–155.

Turksen I. B., Interval-valued strict preference with Zadeh triples, *Fuzzy Sets and Systems*, (1996), 78:183–195.

Wang G. and Li X., Correlation and information energy of interval-valued fuzzy numbers, *Fuzzy Sets and Systems*, (2001), 103:169–175.

Wang G. and Li X., The applications of interval-valued fuzzy numbers and interval distribution numbers, *Fuzzy Sets and Systems*, (1998), 98:331–335.

Wang J. Q., Nie R., Zhang H., and Chen X., New operators on triangular intuitionistic fuzzy numbers and their applications in system fault analysis, *Information Sciences*, (2013), 251:79–95.

Yao J. S., Su J. S., and Shih T. S., Fuzzy system reliability analysis using triangular fuzzy numbers based on statistical data, *Journal of Information Science and Engineering*, (2008), 24:1521–1535.20. Zadeh, L. A., "Fuzzy sets", *Information and Control*, (1965) 8: 338 -353.

Index

Aalen approach, 151
ACMPT-2017, see Analytical and Computational Methods in Probability Theory (ACMPT-2017)
Allan Herschell Company, 129
Amguema-type Arctic cargo ships, 86, 87
Amusement park rides, 121–140
   applications, 128–139
      availability, 138–139
      carousel, 130
      drop zone, 131
      maintainability, 139
      reliability, 131–138
      rotating dragons, 129
      rotating electric train, 128–129
      track with bumper cars, 129–130
   overview, 121–124
   reliability indices, 126–128
      availability, 126–127
      maintainability, 127–128
      mean time to failure (MTTF), 126
   reliability measures, 124–125
Analytical and Computational Methods in Probability Theory (ACMPT-2017), 174
APG-68, F-16 aircraft radar, 25
Applied Stochastic Models in Business and Industry, 178
Arithmetic operations and interval-valued fuzzy numbers, 202
Arnold, Barry C., 164–166
   Jerzy Filus' reply to, 166–168
   rejoinder on joiners by, 168
Arsenic administrative data from Parinacota, 157–159
Arsenic and boron in drinking water, 158
Assembling chain of erasure coding and replication (ASSER), 62
Automatic fault recovery mechanism, 62
Auxiliary functions, 47

Bathtub curve, 125
Bayesian techniques, 19
Bayes theory, 19
Binary-state cloud storage systems, 62
Bivariate distributions, 145, 152, 164
Bivariate lognormal distribution, 156
Bivariate normal distribution, 153
Bivariate pseudoexponential distribution, 162
Bivariate pseudonormal distribution, 147
Bivariate survival function, 151–152, 166, 168
Bone mineral density (BMD) and pseudoexponential models, 155–157
Bumper cars, track with, 123–124, 129–130, 133

Carousel, 124, 130
CBM, see Condition-based maintenance (CBM)
CDF, see Cumulative distribution function (CDF)
CEN, see European Standardization Committee (CEN)
Cloud data reliability management mechanism, 62
Cloud-redundant arrays of independent disk (RAID)
   evaluation methods, 67–74
      disk-level state probability, 67–69
      system-level state probability, 69–74
   numerical results and analysis, 75–79
      disk-level evaluation, 75–76
      system-level evaluation, 77–79
   overview, 61–63
   preliminary models, 64–67
      element-level coverage (ELC) for MSS, 65–66
      multi-state multi-valued decision diagram (MMDD), 66–67
   and reliability modeling, 63–64

215

Cloud storage system, 61, 62
CMs, see Corrective maintenances (CMs)
Complex systems
  application, 24–29
    TWT availability optimization, 26–29
    TWT deterioration, 25–26
  availability computation and optimization, 24
  condition-based model, 20–23
    description, 20–22
    semi-Markov modeling, 22–23
  overview, 17–20
Conditional distributions, 163, 164
Conditional probability of failure, 51
  nr non-repairable, 51
  r repairable, 51
Condition-based maintenance (CBM), 18, 19–20, 30
Condition-based model of complex systems, 20–23
  description, 20–22
  semi-Markov modeling, 22–23
Conditioning method, 146
Content storage and delivery scheme, 62
Continuous Multivariate Distributions, 147
Continuous time Markov chains (CTMCs), 18, 19–20
Copulas, 146–149, 160
Cornell University, 145, 146
Corrective maintenances (CMs), 2, 4, 21
Cost optimal maintenance, 10–14
Cost optimization problem, 10
Cox model, 151
CTMCs, see Continuous time Markov chains (CTMCs)
Cumulative distribution function (CDF), 11, 175

Deterioration process, 19
"Development and Application of Practical and Advanced Dose-Finding Designs," 155
Diesel-generator subsystem, 87, 90–92
DIN 4112 *Fliegende Bauten*, 123
Direct electric propulsion system, 86
Discrete-state continuous-time (DSCT) Markov processes, 84

Discrete-state continuous-time (DSCT) stochastic process, 84
Disk-level state probability evaluation, 67–69, 75–76
Dose exposure levels and pseudoexponential models, 155
Drop zone, 124, 131, 133, 140
DSCT, see Discrete-state continuous-time (DSCT) Markov process; Discrete-state continuous-time (DSCT) stochastic process

ELC, see Element-level coverage (ELC)
Electric motors subsystem, 87, 92–93
Element-level coverage (ELC), 65–66
EM, see Expectation maximization (EM) algorithm
Embedded Markov chain (EMC), 107–108
EN 13814, 123
Erasure coding, 62
European Standardization Committee (CEN), 123
Expectation maximization (EM) algorithm, 110, 118
Exponential distribution, 11, 26, 145, 148

Failure counting processes, 4
Failure rate function, 125
Fault propagation, 18
FDEP, see Functional dependence (FDEP) gate
Filus, Jerzy, 161–162
  beginnings, 144–145
  literature review, 144
  load optimization, 145–146
  Marshall and Olkin model, 149
  rediscovery and generalization of Freund (1961), 145
  reply to Barry C. Arnold's comments, 166–168
  work at Illinois Institute of Technology and alternative to copulas, 146–149
Filus model, 162
Fractal-based cancer and pseudoexponential models, 152–154
Frank, Maurice, 146, 147
Freud, Sigmund, 145

Index

Functional dependence (FDEP) gate, 64
Fuzzy fault tree analysis, 198
Fuzzy reliability of systems and
 interval-valued fuzzy
 numbers, 202–210
 parallel–series system, 206–207
 parallel system, 204–206
 series–parallel system, 207, 210
 series system, 202–203
Fuzzy set, 199
Fuzzy-state assumption, 198

Game-theoretic mechanism, 62
Gamma lifetime distribution, 148
Gaussian process regression (GPR), 19
Gumbel bivariate exponential
 distribution, 153, 162

Happy Park (Kavala), 123–124
Hidden Markov models (HMMs), 105
Hidden Markov renewal chains
 (HMRCs), 106, 107, 108
Hidden Markov renewal models
 description, 107–111
 overview, 105–107
 rate of occurrence of failures
 (ROCOF), 111–118
 full, 111–113
 left, 116–118
 right, 113–115
Hidden semi-Markov chain, 108
Hidden semi-Markov model (HSMM),
 105–106, 119
HMMs, *see* Hidden Markov models
 (HMMs)
HMRCs, *see* Hidden Markov renewal
 chains (HMRCs)
HSMM, *see* Hidden semi-Markov model
 (HSMM)
"Hurricane" ride, 129

IDRGP, *see* Interactive degradation rate
 generic path (IDRGP)
Illinois Institute of Technology (IIT),
 146–149, 146–149
Imperfect fault coverage (IPC), 62, 64, 65
Imperfect maintenance, 2, 18, 21, 26
Independent degradation model, 19
Inherent dependencies, 17

Inhomogeneous Poisson process, 4
Instantaneous availability, 127
Institute of Public Health (ISP), 158
Interactive degradation rate generic
 path (IDRGP), 19
International Conferences on
 Mathematical Methods in
 Reliability (MMR-2017), 174, 178
Interval availability, 127
Interval-valued fuzzy numbers, 197–212
 arithmetic operations on, 202
 definition, 199–201
 fuzzy reliability of systems using,
 202–210
 parallel–series system, 206–207
 parallel system, 204–206
 series–parallel system, 207, 210
 series system, 202–203
 overview, 197–199
Interval-valued trapezoidal fuzzy
 number, 201
Interval-valued triangular fuzzy
 number, 199–200
Intuitionistic fuzzy numbers, 198
Inverse-log-gamma distribution, 161
IPC, *see* Imperfect fault coverage (IPC)
ISO/IEC 2382-14, 127
ISP, *see* Institute of Public Health (ISP)

Jordanova, Pavlina, 160–163
JSM 2016, 155

Kolmogorov differential equations, 85
Kolmogorov forward system, 176
Kotz, Samuel, 147, 149
K-out-of-n: G system, 43–55
 assumptions and state space, 45–46
 long-run distribution, 49–50
 Markov counting process, 52
 measures, 50–51
 availability, 50
 conditional probability of
 failure, 51
 reliability, 51
 model, 46–49
 auxiliary functions, 47
 transition probability matrix,
 47–49
 numerical example, 52–55

overview, 43–45
transient distribution, 49
Kovalenko, I., 173

Laplace–Stieltjes transforms, 174, 178
Laplace transform-based method, 69
Laplace transforms (LTs), 174, 188
Large failures, *see* Type II failures
Lectures from Psychoanalysis, 145
LFRD, *see* Linear failure rate distribution (LFRD)
Likelihood function, 110
Limiting interval availability, 127
Linear failure rate distribution (LFRD), 11
Linear regression concept, 157
Load optimization, 145–146
Loglogistic distribution, 135, 137
Long-run distribution, 49–50
Los, 144
LTs, *see* Laplace transforms (LTs)
Luna Park rides, *see* Amusement park rides
L_z-transform method
  analyzing MSS behavior, 84–85
  multi-state models, 86–101
    calculation reliability indices, 99–101
    diesel-generator subsystem, 90–92
    electric motors subsystem, 92–93
    elements description, 89–90
    for ship's alternative diesel-electric traction drive, 95–98
    for ship's conventional diesel-electric traction drive, 93–95
    systems description, 86–87
  overview, 83–84

Macro-states probabilities, 186–187
Maintainable failures, *see* Type I failures
Maintenance models, 1
Maintenance policy and optimization problem, 2–3
Malfunction, 17–18, 20–21
Marginal distribution, 164
Markov chain, 2
Markov counting process, 52
Markovization method, 176, 194
Markov renewal chain (MRC), 107

Marshall and Olkin model, 149, 162
Marshall-Olkin Weibull models, Pseudo-Weibull and, 157–159
Mathematics Department, UIC, 147
Matrix-analytic methods, 49
Maximum likelihood estimators (MLEs), 106, 110, 118
Mean time to failure (MTTF), 126
MFT, *see* Multi-state fault tree (MFT) model
MIAP, *see* Modified iterative aggregation procedure (MIAP)
Micro-state probabilities, 186
Minor failures, *see* Type I failures
MLEs, *see* Maximum likelihood estimators (MLEs)
MMDD, *see* Multi-state multi-valued decision diagram (MMDD)
MMR-2017, *see* International Conferences on Mathematical Methods in Reliability (MMR-2017)
Modified iterative aggregation procedure (MIAP), 18
Modified Weibull distribution (MWD), 11
Morgenstein–Gumbel bivariate distribution, 157
MRC, *see* Markov renewal chain (MRC)
MSS, *see* Multi-state system (MSS)
MTTF, *see* Mean time to failure (MTTF)
Multiple maintenance models, 2
Multi-state fault tree (MFT) model, 64
Multi-state models, 86–101
  calculation reliability indices, 99–101
  diesel-generator subsystem, 90–92
  electric motors subsystem, 92–93
  elements description, 89–90
  for ship's alternative diesel-electric traction drive, 95–98
  for ship's conventional diesel-electric traction drive, 93–95
  systems description, 86–87
Multi-state multi-valued decision diagram (MMDD), 64, 66–67, 71
Multi-state system (MSS), 44, 62, 65–66, 84, 84–85, 99–100
Multivariate distributions, 164, 166

Index

Multivariate pseudonormal distribution, 147
MWD, see Modified Weibull distribution (MWD)

Non-maintainable failures, see Type II failures
Non-negative probability distributions, 44
non-repairable failure, 51

One-step probability matrix, 23
Operational availability, see Interval availability

Pakistan Journal of Statistics (PJS), 147
PAO, see Polyalphaolefins (PAO)
Parallel–series system, 206–207, 212
Parallel system, 204–206, 212
Parameter dependence method, 146, 163
Parameter dependence-related constructions *vs.* stochastic dependence, 149–152
Pareto diagram and theory, 131–132, 133
Pareto model, 154
Partial differential equations, 176
Periodic maintenance models, 2
Phase-type distribution (PH), 44
Piasecki, Stanislaw, 144
PJS, see Pakistan Journal of Statistics (PJS)
PM, see Preventive maintenance (PM)
Point availability, see Instantaneous availability
Poisson processes, 163
Poisson-type failures, 19
Polish Academy of Sciences, see System Research Institute
Polyalphaolefins (PAO), 25, 26
Possibility assumption, 198
Predictive maintenance, 18, 20
Preventive maintenance (PM), 1, 2, 3, 11, 12, 18
Proactive replica checking approach, 62
Probability density function, 161, 175
Probability laws, 162
Proschan, F., 146
Pseudoaffine transformation, 147
Pseudoexponential models, in medicine, 152–157
    bone mineral density, 155–157
    dose finding studies, 155
    fractal hypothesis for cancer, 152–154
Pseudo-lognormal distribution, 156
Pseudonormal distribution, 148
Pseudo-Weibull and Marshall-Olkin Weibull models, 157–159
Psychoanalysis: Evolution and Development, 145

Quasi-stationary probabilities (q.s.p.), 174, 188–190, 193–194

Rate of occurrence of failures (ROCOF), 106–107, 111–119
    full, 111–113
    left, 116–118
    right, 113–115
Rayleigh distribution (RD), 11
Reduced modified Weibull distribution (RMWD), 11, 12
Redundant systems, 43
Regional Ministerial Secretary (SEREMI), 158
Regression technique, 160
Reliability, of rides
    indices, 126–128
    availability, 126–127
    maintainability, 127–128
    mean time to failure (MTTF), 126
    measures, 124–125
    time distributions determination, 133
    TM variable, 133
    TTF variable, 135–136
    TTR variable, 136
    TTR1 variable, 137–138
Renewable systems, 173–194
    overview and motivation, 173–174
    problem setting and notation, 175–176
    quasi-stationary probabilities, 188–190
    reliability function study, 176–179
        calculation, 176–178
        sensitivity analysis, 178–179
    sensitivity analysis, 191–193
        full repair, 192–193
        partial repair, 191–192
        quasi-stationary probabilities, 193

stationary probabilities, 180–187
  full repair, 184–187
  partial repair, 180–184
Repairable failure, 51
RMWD, see Reduced modified Weibull distribution (RMWD)
ROCOF, see Rate of occurrence of failures (ROCOF)
Rotating dragons, 123, 129, 133
Rotating electric train, 123, 128–129
Rykov, V., 174
Ryll-Nardzewski, 144

Semi-Markov decision process, 19
Semi-Markov kernel (SMK), 107
Semi-Markov model (SMM), 20, 22–23, 106, 119
Semi-Markov process (SMP), 19, 23
Sensitivity analysis, 191–193
  full repair, 192–193
  partial repair, 191–192
  quasi-stationary probabilities, 193
Sequential imperfect PM model, 2
SEREMI, see Regional Ministerial Secretary (SEREMI)
Series–parallel system, 207, 210, 212
Series system, 202–203, 212
Sevast'yanov, B., 173
Ship's diesel-electric traction drive
  alternative, 95–98
  calculation reliability indices, 99–101
  conventional, 93–95
Sklar, Abe, 146, 148
SMK, see Semi-Markov kernel (SMK)
SMM, see Semi-Markov model (SMM)
SMP, see Semi-Markov process (SMP)
Stationary probabilities, 180–187
  full repair, 184–187
  partial repair, 180–184
Steady-state availability, 127
Steady-state probabilities (s.s.p.), 174, 182–184, 191–193
Stochastic dependence, parameter dependence-related constructions vs., 149–152
"Stochastic Dependence and Related Topics," 144

Stochastic dependencies, 17
Stochastic processes, 2, 4
Stochastic reaction, 153
System-level state probability evaluation, 69–74
  controller with imperfect coverage, 78–79
  controller with perfect coverage, 77–78
  covered failure, 72–73
  degraded, 73–74
  good state probability, 70–72
  perfect controller, 77
  uncovered failure, 74
System reliability function, 176–179
  calculation, 176–178
  sensitivity analysis, 178–179
System Research Institute, 144
System with two failure types, 1–14
  cost optimal maintenance, 10–14
  maintenance policy and optimization problem, 2–3
  modeling, 4–10
  overview, 1–2

Taylor, H. M., 145
Taylor expansion, 192
Ternary decision diagrams (TDDs), 71, 73
Thermal conductivity, 25
Thompson, Clara, 145
Time-dependent stochastic process, 19
Time distributions determination, 133
Time interval maintenance, 18
Time of maintenance (TM), 133
Time till failure (TTF), 132, 135–136
Time till repair (TTR), 136–138
TM, see Time of maintenance (TM)
Traditional reliability theory, 44
Transient distribution, 49
Transition probability matrix, 47–49, 56–57, 58
Traveling wave tubes (TWTs), 24–25
  availability optimization, 26–29
  deterioration, 25–26
Triangular fuzzy numbers, 198
TTF, see Time till failure (TTF)
TTR, see Time till repair (TTR)

Index

Two-dimensional Markov process, 180, 185
Two-level hierarchical methods, 62
TWTs, *see* Traveling wave tubes (TWTs)
Type I failures, 1–2, 3, 4, 8–10
Type II failures, 1, 2, 3, 4, 9

Universal generating function (UGF), 84

Vector Markov process, 46

Warsaw University, 144
Weibull distribution, 135, 137